FATIGUE AS A DESIGN CRITERION

FATIGUE AS A DESIGN CRITERION

Terance V. Duggan and James Byrne

Department of Mechanical Engineering and
Naval Architecture, Portsmouth Polytechnic

First published 1977 by
THE MACMILLAN PRESS LTD
London and Basingstoke
Associated companies in New York Dublin
Melbourne Johannesburg and Madras

ISBN 0 333 21488 9

Typeset in IBM 10/11 Press Roman by
Santype International
(Coldtype Division)
Salisbury, Wilts.

Printed in Great Britain by
Unwin Brothers Limited
The Gresham Press,
Old Woking, Surrey

Contents

Preface

The majority of research studies in the field of metal fatigue or progressive fracture have been mainly concerned with fundamental investigations into the basic mechanism, or have concentrated on the specific assessment of various factors. Consequently, although the literature available dealing with the subject is immense and continues to increase at a tremendous rate, it remains a fact that the application of fatigue data to design situations has not received such great attention, and it is still common practice to carry out fatigue tests on actual components.

The basic objective of this book is that of establishing methodical procedures for the application of fatigue data to design. In so doing, the mechanical behaviour of materials, the factors which influence such behaviour and the application to real components are discussed.

In assessing the fatigue integrity of components, due consideration must be given to stress—strain analysis, crack formation, fatigue crack growth and final fracture. All these aspects are dealt with in detail, and some new research data and original theory are also included. The work is based on experience gained over a considerable period of time, both as a research activity, and also in teaching the subject at both under- and post-graduate level to engineering students.

This book is not intended to supersede the various excellent texts dealing with particular aspects of fatigue. However, every effort has been made to ensure adequate coverage of both high and low cycle fatigue; multiaxial stress systems subjected to cyclic loading; simple creep-fatigue considerations; cyclic material behaviour; metallurgical aspects of fatigue; the assessment of crack formation life; fracture mechanics and its application to fatigue crack growth and fast fracture; and the influence of environmental factors. Each chapter includes an extensive and comprehensive list of references, and the reader who studies this work will be in a strong position to read the most current research papers associated with fracture and fatigue.

It is hoped that the book will appeal to research workers, stress analysts, design engineers and metallurgists. Additionally, the book has been written to satisfy the requirements of most mechanical engineering degrees and diplomas which include the mechanical behaviour of materials, and it should provide a useful background text for adoption in many University and Polytechnic Departments. Indeed, in supervising final year student projects in this field over the past decade, the need for a single comprehensive but simple text has become apparent, and we sincerely hope that this book will go at least part way to satisfy this need.

It should be mentioned that the material presented here has been used most successfully, in note form, as the basis for short post-graduate courses operated in the Department of Mechanical Engineering at Portsmouth Polytechnic. Consequently, we would like to thank our colleagues in the Mechanical Behaviour of Materials Group for their valuable assistance in both direct and indirect ways, as well as our industrial collaborators, particularly our friends at Rolls Royce Ltd. associated with mechanical integrity and materials engineering. Finally, we wish to record our thanks to the publishers, and all associated with the production of the book, for their care, interest and cooperation throughout. Any criticisms of the work should, of course, be directed to the authors!

T.V.D.
J.B.

Portsmouth, 1977

Notation

A	$\begin{cases} \text{material constant} \\ \text{cross sectional area} \end{cases}$
$\left.\begin{array}{l} A_1 \\ A_2 \\ A_3 \end{array}\right\}$	constant coefficients
a	$\begin{cases} \text{half crack length for central crack} \\ \text{crack length for edge crack} \\ \text{half length of minor axis of ellipse} \\ \text{hole radius} \\ \text{Neuber material constant} \end{cases}$
a_c	critical crack length
B	$\begin{cases} \text{material constant} \\ \text{specimen thickness} \end{cases}$
b	$\begin{cases} \text{material constant} \\ \text{half length of major axis of ellipse} \end{cases}$
C	$\begin{cases} \text{material constant} \\ \text{creep rate} \end{cases}$
C_e	strain range corresponding to elastic intercept for one cycle
C_L	factor to allow for type of loading
C_p	$\begin{cases} \text{strain range corresponding to plastic intercept for one cycle} \\ \text{condition for non-propagating crack} \end{cases}$
C_s	factor to allow for size effects
$\left.\begin{array}{l} C_1 \\ C_2 \\ C_3 \end{array}\right\}$	creep rates in directions of principal stress material constants
$\left.\begin{array}{l} c \\ c_1 \\ c_2 \\ c_3 \end{array}\right\}$	combined material and geometric constants
D	$\begin{cases} \text{damage factor} \\ \text{logarithmic ductility} \end{cases}$
da/dN	fatigue crack growth rate
E	elastic modulus
E_s	secant modulus
F	$\begin{cases} \text{material constant used to describe stress–strain behaviour} \\ \text{constraint factor} \end{cases}$
f	$\begin{cases} \text{function} \\ \text{frequency} \end{cases}$
G	elastic modulus in shear
$\mathscr{G}$	strain energy release rate

g	material constant
$\left.\begin{array}{l} I_1 \\ I_2 \\ I_3 \end{array}\right\}$	stress invariants
K_c	critical stress intensity factor
K_f	fatigue strength reduction factor
K_{fs}	fatigue strength reduction factor in shear
K_s	factor to allow for surface finish effects
K_t	theoretical or geometric stress concentration factor
K_{ts}	theoretical or geometric stress concentration factor in shear
K_{max}	maximum stress intensity factor
K_{min}	minimum stress intensity factor
K_ϵ	strain concentration factor
K_σ	stress concentration factor
K_I	stress intensity factor for mode I opening
K_{II}	stress intensity factor for mode II opening
K_{III}	stress intensity factor for mode III opening
K_{IC}	plane strain fracture toughness
K_{ISCC}	threshold stress intensity in specific corrodent
k	material constant
M	bending moment
m	material constant
N	number of applied cycles
N_f	number of cycles to failure
N_c	number of cycles to give critical condition
n	material constant
p	numerical exponent
Q	flaw shape parameter
q	$\left\{\begin{array}{l} \text{numerical exponent} \\ \text{notch sensitivity index} \end{array}\right.$
q_s	notch sensitivity index in shear
R	bulk stress ratio
RA	reduction in area
RF	reserve factor
RF_a	reserve factor for alternating stress
RF_m	reserve factor for mean stress
r	notch radius
r_p	plastic zone radius
r_ϵ	localised strain ratio
r_σ	localised stress ratio
S	surface energy
S_a	intrinsic fatigue strength corresponding to any number of cycles for completely reversed loading
S_a'	component fatigue strength corresponding to any number of cycles for completely reversed loading
S_e	intrinsic fatigue or endurance limit for completely reversed loading
S_e'	component fatigue or endurance limit for completely reversed loading
S_c	critical strength of material or component

S_f	fatigue strength coefficient
S_p	monotonic yield or proof strength
S_p'	cyclic yield or proof strength
S_{sa}	intrinsic fatigue strength in shear corresponding to any number of cycles for completely reversed loading
S_{sa}'	component fatigue strength in shear corresponding to any number of cycles for completely reversed loading
S_{sp}	monotonic yield strength in shear
S_{su}	ultimate strength in shear
S_u	ultimate tensile strength
S_w	working strength
T	failure time at particular load condition
t	$\left\{ \begin{array}{l} \text{thickness} \\ \text{time duration of applied loading} \end{array} \right.$
U	strain energy
W	specimen depth
x	linear co-ordinate
Y	compliance function
Y_0	modified compliance function
y	linear co-ordinate
z	linear co-ordinate
α	material constant
α_1	slope of plastic strain with cycles on logarithmic plot
α_2	slope of elastic strain with cycles on logarithmic plot
β	angle measured around crack front
γ	shear strain
γ_s	surface strain energy per unit area
γ_p	plastic work factor per unit area
ΔK	stress intensity range
ΔK_I	stress intensity range for mode I opening
ΔK_{ITH}	stress intensity range below which fatigue cracks remain dormant
ΔK_{I0}	threshold stress intensity range
$\Delta \epsilon$	strain range
$\Delta \epsilon_e$	elastic strain range
$\Delta \epsilon_a$	endurance strain range
$\Delta \epsilon_{e0}$	elastic strain range with zero mean stress
$\Delta \epsilon_{L0}$	strain range equivalent of endurance or fatigue limit for zero mean stress
$\Delta \epsilon_T$	total strain range
$\Delta \sigma$	localised stress range
$\Delta \sigma_0$	nominal or bulk stress range
δ	critical depth of material required to produce failure
ϵ	strain amplitude
$\left. \begin{array}{l} \epsilon_1 \\ \epsilon_2 \\ \epsilon_3 \end{array} \right\}$	principal strains
ϵ_c	strain amplitude corresponding to critical condition
ϵ_f	fracture strain

ϵ_f'	fatigue ductility coefficient
ϵ_{end}	endurance or fatigue limit strain equivalent
ϵ_m	mean strain
ϵ_{max}	maximum strain amplitude
ϵ_{min}	minimum strain amplitude
ϵ_0	nominal strain amplitude
ϵ_{yp}	yield strain
θ	polar co-ordinate
ν	Poisson's ratio
ν_e	Poisson's ratio in elastic region
ν_p	Poisson's ratio in plastic region
ν'	modified Poisson's ratio to allow for some degree of plasticity
ρ	notch radius
σ	direct stress
σ_{alt}	alternating component of direct stress
σ_{alte}	equivalent alternating component of direct stresses
σ_e	equivalent stress amplitude
σ_f	fracture stress
σ_m	mean component of direct stress
σ_{me}	equivalent mean component of direct stresses
σ_{max}	maximum stress amplitude
σ_{min}	minimum stress amplitude
σ_0	nominal or bulk stress amplitude
σ_s	equivalent static stress replacing cyclic stresses
$\left.\begin{array}{c}\sigma_1\\\sigma_2\\\sigma_3\end{array}\right\}$	principal stress amplitudes
τ	shear stress
τ_{alt}	alternating component of shear stress
τ_m	mean component of shear stresses
τ_{max}	maximum shear stress amplitude
τ_{min}	minimum shear stress amplitude
$\left.\begin{array}{c}\tau_{12}\\\tau_{13}\\\tau_{23}\end{array}\right\}$	maximum shear stress amplitudes
ϕ	$\left\{\begin{array}{l}\text{plasticity modulus}\\\text{elliptic integral}\end{array}\right.$

1

Factors Affecting Fatigue Behaviour

1.1 Introduction

The presently used approach for estimating fatigue strength or cyclic life of a component leaves much to be desired, and there is a definite need for the establishment of reliable methods for assessing fatigue integrity. Any such methods should utilise the basic material data which may be obtained by testing simple specimens in the laboratory either under constant load (stress), constant strain (displacement), or possibly a constant product of stress and strain. Such tests should, for convenience and simplicity, preferably be carried out on plain specimens, although the possible advantages of testing some type of notched specimen or simulated component must also be considered.

Fatigue behaviour is dependent upon many factors and, apart from testing actual components under service conditions, it is not possible to include all likely influencing factors in laboratory tests. Consequently when applying fatigue data obtained in the laboratory to the design of actual components, due account must be taken of those factors likely to affect fatigue behaviour. If one considers an actual component and compares this with a simple fatigue test piece, the differences are at once apparent. For a variety of obvious reasons it is not always possible to carry out actual component testing, and the designer must content himself with basic material data available, and use a method of analysis which utilises this information.

Experimental data indicates that the same factors which affect the tensile strength of a material also affect the fatigue strength (the reverse of this statement is not necessarily true). As a consequence, attempts have been made to relate the fatigue strength with tensile strength[1,2]. It must be emphasised that this is not a practice to be generally recommended, but in the absence of more reliable information, it is one which the designer might nevertheless find useful in many situations involving initial feasibility studies, where basic fatigue data is not readily available. In any event, basic or intrinsic fatigue data is obtained using plain test specimens manufactured and controlled to a high degree of accuracy and surface finish, of a standard size, and usually subjected to a completely reversed or repeated one-dimensional load pattern of constant amplitude in a non-hostile environment (relatively speaking). These conditions are quite remote from real design conditions, and at the best the test specimens reflect only those factors which might be considered fundamental to the material at room temperature. This enables the

1

intrinsic fatigue strength to be used for purposes of material comparison, and whilst this is of some value, it is limited as far as application to designed components are concerned.

One possible approach to the application of fatigue data to design, which has not yet been properly explored[3] would seem to be that of determining the monotonic and cyclic strain distributions for the component and attempting to relate such distributions to nucleation, crack propagation and rate of crack growth. Ideally, the monotonic and cyclic strain distributions should be obtained by theoretical considerations using known material behaviour relationships. It is the view of the authors that the monotonic strain distribution will be of considerably less consequence than that of the cyclic strain distribution, since cyclic strain hardening or softening will undoubtedly influence the fatigue process[3]. Any theoretical approach to the determination of strain distributions in a component requires the satisfaction of three basic relationships, namely:

(1) the equilibrium equations;
(2) the compatibility equations; and
(3) the material stress–strain relationship.

When loading conditions are such that only elastic strains are evident, the material behaviour relationship is represented by Hooke's law, and provided that the equilibrium and compatibility equations can be satisfied, the stress and strain distributions can be obtained. However, if the loading is such that local yielding occurs in the component or structure, plastic as well as elastic strains will be present, and the material behaviour relationship is no longer linear.

1.2 High Cycle and Low Cycle Fatigue

The prediction of the intrinsic fatigue curve for a given material provides the designer with the basic material data required in order to design against failure under dynamic loading. At this stage, the intrinsic fatigue curve will be defined as that curve obtained in the laboratory from tests on plain specimens manufactured and polished to a high degree of accuracy, and subjected to completely reversed one-dimensional stresses. The way in which this basic or intrinsic fatigue curve needs to be modified for application to design situations requires the consideration of the factors which affect fatigue strength, but which are generally not present in the laboratory test pieces.

Fatigue problems can, to a large extent, be conveniently divided into two categories[4], namely those which involve strain (or deformation) cycling, and those which involve load (or nominal stress) cycling. It would, therefore, not appear unreasonable that in order to obtain intrinsic fatigue data for particular materials, the method of testing plain specimens in the laboratory should correspond to that category of cycling which the component is likely to be subjected to in service; this apparent logic is, unfortunately, somewhat superficial. For the present purpose, it is sufficient to indicate that fatigue curves may be obtained by testing plain specimens in the laboratory under conditions of either constant strain (or deformation) cycling or, by the more conventional method in the past of constant load (or, loosely speaking, stress) cycling. Provided that the stress levels do not significantly exceed the macroscopic yield strength of the material, the two types of curves will be more or less identical. On the other hand, where high strain and consequently significant macroscopic plasticity is involved, the relationship between stress and strain is no longer linear, and is further complicated under cyclic loading conditions by the cyclic strain hardening or strain softening characteristics exhibited by the material[5]. It therefore becomes necessary to distinguish between high strain low cycle fatigue (LCF) and high cycle fatigue (HCF), the former usually being obtained by testing specimens under conditions of constant strain (or deformation) and the latter under conditions of constant load.

High cycle fatigue, whilst inferring that a large number of cycles is required to cause failure, refers to those combinations of stress (or strain) and cycles during which macroscopic plasticity or yielding does not occur. Since under these conditions the usual elastic relationship between stress and strain applies, HCF curves are usually represented by stress amplitude (S_a) as the ordinate, and number of cycles to failure (N_f) as the abscissa, as indicated in figure 1.1.

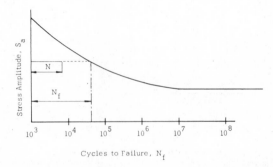

Figure 1.1 *This fatigue curve for HCF region was obtained by testing several identical specimens at different stress levels and fitting the best curve to the test data*

Low cycle fatigue whilst inferring that only a small number of cycles (relatively speaking) is required to cause failure, refers to those combinations of strain and number of cycles during which considerable macroscopic plasticity occurs. Under these circumstances the linear relationship between stress and strain is invalid, and LCF curves are usually represented by strain range $(\Delta \epsilon)$ as the ordinate and number of cycles to failure (N_f) as the abscissa as indicated in figure 1.2. The terminology used in the plastic region is illustrated by the hysteresis loop shown in figure 1.3. LCF as a criterion of failure is possible in many dynamic situations where the stresses accompanying cyclic loading exceed the macroscopic yield strength, typical situations being found in nuclear pressure vessels, aero-engines, modern steam turbines, and so on.

Whether in the HCF or LCF region, the usual way of determining the intrinsic fatigue strength of a material is experimentally. This is achieved by subjecting test specimens to repeated loads or strains of specified amplitudes or ranges, and determining the number of cycles required to produce failure. Such results, when plotted graphically, give the curves already mentioned (figures 1.1 and 1.2).

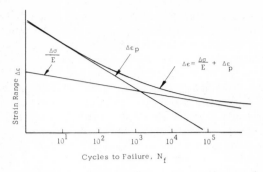

Figure 1.2 *In the LCF region total strain is asymptotic to the plastic strain component but in the HCF region it is asymptotic to the elastic strain component*

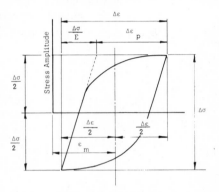

Figure 1.3 *For constant strain range the corresponding stress range will increase (cyclically strain hardening) or decrease (cyclically strain softening) with continued cycling until a stable loop is obtained*

The frequently used method of obtaining an intrinsic fatigue curve is to test a few apparently identical specimens at each load or strain level, and then to visually fit a smooth curve through these points. It is now generally recognised that scatter in fatigue life is an inherent property of the material and must be considered as a stochastic quantity[6]. There would also appear to be general agreement that the main reasons for scatter in fatigue life is due to the critical dependence of fatigue nucleation on the microscopic properties of the material. This is borne out by the fact that in the LCF region, where crack propagation takes up the majority of cyclic life, the amount of scatter is considerably reduced. In any event, the method of visually fitting the so-called best curve through a few experimentally determined points leaves much to be desired. At the same time, for economic and other reasons, it is not always possible to obtain fatigue data using a very large number of specimens. Further, however many specimens are used to obtain the experimental data sufficient to predict an intrinsic fatigue curve, the question arises concerning how this data may best be interpreted, particularly for use in design.

The statistical nature of fatigue has been the subject of numerous investigations[7-11], and there is evidence to suggest that there is a possibility of a hump or discontinuity in the fatigue curve for some materials; although different interpretations have been placed upon this latter point[12], the practical significance is not fully understood. The statistical nature of fatigue, particularly in the HCF region, is of such importance that it must constantly be remembered and accounted for in one way or another.

It should also be mentioned that in the case of components containing discontinuities or changes in geometry, the scatter is more likely to be due to directional properties, differing metallurgical and heat treatment conditions, methods of fabrication, and so on, than to the critical dependence of fatigue crack formation on the maximum localised stress and the microscopic properties of the material. The reason for this is simply that the crack formation period is considerably reduced and a larger number of cycles, as a proportion of the total life, is spent in crack propagation.

Accepting that probability fatigue curves[7] for specific materials can be obtained by fatigue testing samples of apparently identical test specimens in the laboratory, the question must still be answered concerning how this data can be applied realistically to design situations. In other words, of what value to the designer is the basic or intrinsic fatigue curve obtained in the laboratory? This question is one of vital importance, and is the major subject of interest in the present text. Of course, one must first of all ask the question regarding the type of fatigue curve to be used, i.e. whether this should be obtained under conditions of constant load (stress) or constant strain (deflection), and under what conditions of loadings, i.e. axial, bending, torsional or perhaps combined loadings? The ideal situation, if one can imagine it, would be that of predicting the fatigue integrity for an actual component, such that combined loadings, stress or strain concentrations and so on, can easily be incorporated, based on a non-destructive method of testing, or perhaps from data obtained from the simple static tensile test on the material of interest. The ideal, however, is hardly likely to be realised.

In any fatigue analysis for a component, whether this be for the LCF or HCF region, it will be necessary to take account of the many influencing factors which effect fatigue behaviour, amongst the major ones of which are:

(1) type and nature of loading;
(2) size of component and stress or strain distribution;
(3) surface finish and directional properties;
(4) stress or strain concentrations;
(5) mean stress or strain;
(6) environmental effects;
(7) metallurgical factors and material properties;
(8) strain rate and frequency effects.

1.3 Designing Against High Cycle Fatigue

It is useful to consider the presently used engineering approach to designing against fatigue in the intermediate and high cycle region. The upper curve in figure 1.4 represents the intrinsic fatigue strength for the material, i.e. that curve which is usually obtained in the laboratory from standard tests on apparently identical test specimens having a smooth polished surface and of a standard size (typically about 8 mm diameter). The test may be performed at a particular temperature and any other environmental conditions of interest. Further, the tests may be such that superimposed mean and alternating loads are applied, but more usually the condition of completely reversed constant amplitude loading is used. Before this intrinsic fatigue curve can be applied to a real component, it must be modified to

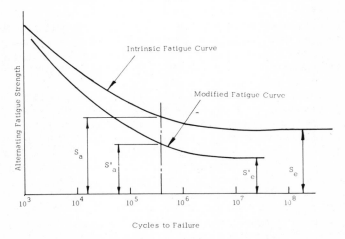

Figure 1.4 *Intrinsic and component fatigue curves*

account for those factors not included but present in the component. Simplifying the situation for purposes of illustration, the fatigue strength of a component subjected to a one-dimensional constant amplitude completely reversed load cycling conditions, might be represented by the equation

$$S_a' = \frac{S_a}{K_f} K_s C_s C_L \qquad (1.1)$$

where S_a' is the modified fatigue strength of the component corresponding to a particular life; S_a is the intrinsic fatigue strength for the material at a temperature and set of environmental conditions of interest, for the same life; K_s, C_s, C_L and K_f represent the surface finish, size, loading and fatigue strength reduction factors respectively, at the particular life under consideration. The lower curve in figure 1.4 represents the modified fatigue curve for the component, obtained by the application of the factors indicated above to the intrinsic fatigue curve.

The above approach does not account for the effect of mean stresses, nor for the fact that in service, components will be subjected to combined dynamic loading. Further complications also arise when the loading is not of constant amplitude. These additional complicating factors will be dealt with in some detail at a later stage, but first of all it is desirable to be able to associate numerical values with those factors included in equation (1.1).

1.4 Effect of Type of Loading

At the present time, only one-dimensional loading systems will be considered, and since most fatigue test data is obtained under conditions of rotating bending, it is convenient to relate other forms of one-dimensional loading to this. In practice, three major types of loading will be encountered, namely bending, axial and torsional. It has been suggested[13] that the type of loading can be conveniently accounted for by the use of a *load correction factor*, designated C_L. For the condition of rotating bending, since this coincides with the type of test generally used in obtaining the basic data, C_L will be equal to unity. Thus, by definition, the load correction factor will be used to account for those situations where the loading is other than rotating bending. It should, perhaps, be mentioned that due to the statistical nature of fatigue itself, it is to be expected that there would be some difference between rotating bending and reversed bending. This is due to the fact that under conditions of rotating bending every element on the critical diameter of the surface will be subjected to the maximum stress level, whereas for reversed bending, where the specimen is bent first in one direction and then by the same amount in the opposite direction, only a small region on the outer surface will experience the maximum stress levels. Consequently, it is possible in the latter case for the weakest points to escape the maximum stress levels, and on a probability basis, the fatigue strength under reversed bending should be somewhat greater than under rotating bending. However, this difference is, in fact, quite small, being in the region of no more than 5 per cent at the endurance limit for steels. From a design point of view, since the other influencing factors are of much greater significance, it is usually assumed that the fatigue strengths under rotating and reversed bending are the same. For most steels, the endurance limit is approximately 0.5 of the ultimate tensile strength, S_u.

The endurance or fatigue limit for direct or so-called push—pull stress conditions is, in general for most materials, somewhat less than the endurance limit obtained under rotating bending. Although the difference is not fully understood, it is undoubtedly connected with the two factors. In the first instance it is difficult to apply axial loads without at least some eccentricity, and this eccentricity will give rise to bending as well as direct stress. Secondly, with true axial loading the whole volume of the critical region will be subjected to the same maximum stress level, i.e. a condition of zero stress gradient is obtained. On this basis, because of the statistical distribution of material strength on a microscopic basis[14], it would be expected that the fatigue strength for axial or push—pull loading would be less than for rotating bending. For most materials the endurance limit for direct stress may be taken as being about 15 per cent less than for rotating bending, although it has been suggested[15] that in some instances it might be as high as 30 per cent. For general use, however, it is suggested that the difference of 15 per cent be utilised, which amounts to using a load correction factor of $C_L = 0.85$. At 10^3 c, for steels, experimental data indicates that the fatigue strength under axial loading conditions is about $0.75S_u$. Thus if these two points are plotted on log-log co-ordinates, a linear relationship may be assumed, and the points connected by a straight line.

Under reversed torsional loading it would obviously be expected that the fatigue strength would be different from that obtained under either rotating bending or axial loading. Of course, the type of stress under torsional loading will be different from the two previous cases mentioned, and it is likely, certainly for ductile materials, that the criterion of failure will be the shear stress. Fatigue tests on steel specimens in torsion seem to indicate that the endurance limit is approximately 60 per cent of the endurance limit in bending. More correctly, on a theoretical basis, this factor may be taken as 58 per cent, in accordance with the von Mises criterion. Thus, for reversed torsional loading, the load correction factor C_L is taken as 0.58. At the 10^3c point, it is usually assumed that the fatigue strength in reversed torsion is about $0.9S_{su}$. In the absence of other information Dolan[16] suggests using the approximation that $S_{su} = 0.83S_u$, although this value is the average of tests made on 35 steels having a range of values between 0.68 and 0.96. It would seem, therefore, more reasonable and conservative, in the absence of specific information, to assume that $S_{su} = 0.75S_u$. Again, if these two points, i.e. the fatigue strengths corresponding to 10^3 and 10^6, are plotted on a logarithmic plot, they may be joined by a straight line.

It should be emphasised that the generalised fatigue curves for the three different types of loading are approximate for steels, and should be used only in the absence of more precise information.

1.5 Effect of Size and Stress Gradient

The fatigue strength of material under conditions of bending and torsion has been observed to vary with size. It is thought that this is connected with the stress gradient, since no noticeable size effect is apparent under direct loading. Under static loading in torsion, results obtained by Morrison[17] indicate that a considerable size effect is present for mild steel; similarly, Morkovin and Moore[18] have obtained results which indicate a considerable size effect in bending fatigue. Further tests carried out by Phillips and Heywood[19] on mild steel indicate the absence of any significant size effect under alternating tension. This evidence would appear to confirm the fact that stress gradient is responsible for any size effect.

It has been suggested[17,20] that yielding cannot readily occur in an individual crystal surrounded by unyielding material, but only in a number of crystals which occupy sufficient volume to allow for the readjustment. It is further suggested that size effect under both static and fatigue conditions is due to the fact that failure must occur in a finite volume of material. Several theories have been suggested to account for the size effect on fatigue strength, and these are reviewed by Heywood[21].

A simple representation of the effect of size and stress gradient is illustrated in figure 1.5, which indicates two specimens of circular cross-section, each being subjected to the same maximum fibre stress. However, it is clear that due to the steeper stress gradient, the smaller of the two specimens will be subjected to a smaller average stress value over a depth δ than will the larger specimen for the same depth δ. If this depth δ is taken to represent the critical volume of material required to be subjected to a critical stress before failure occurs, then it is obvious that failure is more likely to occur in the larger of the two specimens, and this concept is consistent with the view previously presented.

Since most fatigue data available in the past has been obtained on standard test specimens, typically in the region of about 8 mm (approximately 0.3 in) diameter,

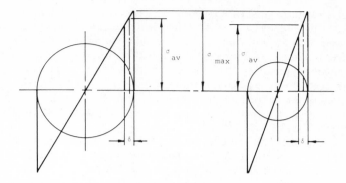

Figure 1.5 *A simple representation of the effect of size and stress gradient*

considerable caution is necessary when applying such data to large components. For components smaller than the test specimen (for example, wire), endurance limits obtained are somewhat higher than those obtained for the test specimens; for larger sizes, the fatigue strength decreases. As a first approximation, a useful design approach is to use the idea of a size correction factor, as suggested by Lipson and Juvinall[22]. Experimental data indicates that up to about 13 mm (0.5 in) diameter the strength decreases by approximately 15 per cent, after which it appears to vary little up to about 50 mm (2 in) diameter. Thus, representing this information in terms of a size correction factor C_s, figure 1.6 illustrates the effect of size on the

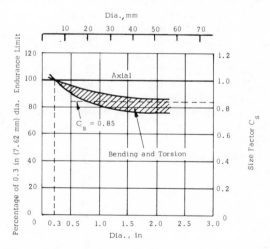

Figure 1.6 *Effect of size on endurance limits for steels (Courtesy Macmillan)*

endurance limit for steels for the range mentioned, with the size correction factor indicated. It is recommended that for specimens between about 13 and 50 mm (0.5 and 2 in), the value for C_s be taken as 0.85 for bending and torsion, and unity for axial loading. Somewhat lower endurance strengths are possible for larger components, but since large component testing is costly and requires special test rigs, there is little direct experimental evidence. Of the data which has been obtained, that of Coyle and Watson[23] is worthy of special mention.

The effect of size at 10^3c for steel is considerably less than it is at the endurance limit, and the usual practice is to consider the effect of size to be insignificant at 10^3c. When correcting the generalised fatigue curve to allow for the size effect, it is usual to apply the value for C_s at the endurance limit, to leave the fatigue strength at 10^3c unaffected, and then to join these two points by a straight line when plotted logarithmically.

The discussion presented here is intended to suggest a simple and useful approach for design purposes. A more fundamental approach would need to consider possible relationships between stress distribution, the material stress—strain relationship, the magnitude of the applied stress and the ratio of plastically to elastically strained material, this latter possibly on a microstructural basis.

1.6 Surface Finish and Directional Properties

Fatigue test specimens are carefully controlled to a high degree of accuracy and usually have a polished finish. In practice, a real component from a machine or structure will generally have a surface finish different from that of the test specimen, and this might have the effect of considerably reducing the fatigue strength. The surface finish of a component will depend upon its method of production, and three effects may be introduced. In addition to the surface roughness or topography, it is likely that residual surface stresses and surface strain hardening will be introduced, but for the present purpose the three contributing factors will be grouped together. Because a fatigue failure generally originates at the surface, it will be evident that the surface finish may be of great importance when designing components to withstand fluctuating loads. Several investigators[24-26] have studied the effect of surface finish on fatigue strength, and there is evidence that, for steels, the higher the tensile strength the more critical is the surface finish. One might suspect that this is because steels of higher tensile strength generally have a finer grain structure, thus making them more *notch sensitive* than steels of lower tensile strength.

Lipson and Juvinall[22] suggest the classification of surface finish into five broad categories, namely:

(1) Polished;
(2) ground;
(3) machined;
(4) hot rolled; and
(5) as forged.

For the purpose of application in design, it is convenient to utilise a surface finish factor, designated K_s, and this factor is dependent upon both the tensile strength and the type of surface finish. Figure 1.7 enables an estimate of K_s to be obtained. Thus, to take account of the surface finish on steel components, the

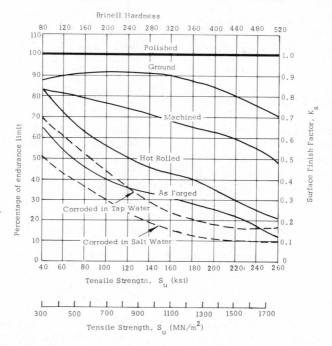

Figure 1.7 *Reduction of endurance limit due to surface finish for steels (Courtesy Macmillan)*

surface finish factor K_s can be determined from figure 1.7 and the endurance limit is multiplied by this value. The effect at 10^3c is usually insignificant and may reasonably be neglected. The variation in fatigue strength between these two points when plotted logarithmically is linear.

The above comments apply specifically to steels, and data are not presently available for assessing the effect of surface finish in other metals. It is common practice to ignore the surface finish effect for metals other than steels and, apart from metals which are not notch sensitive, this presumably leads to some error on the optimistic side. In view of this, the designer might well be inclined to make some allowance for surface finish on metals other than steels by increasing the value for the general reserve factor.

Because the condition of the surface can have a profound effect on fatigue strength, attempts have been made to increase the fatigue strength by suitable surface treatments. It is well known, for example, that the introduction of favourable compressive stresses into the surface, such as by shot peening and surface rolling, can considerably increase the fatigue strength[27-30]. The fatigue strength may also be improved by suitable heat treatment which causes surface hardening, for example by carburising, nitriding, flame hardening or induction hardening[22,24,31-33]

It is also important to note that the effect of inclusions and the preferred orientation of the grains brought about by working, may introduce directional properties into a metal. This anisotropy is very important when the design of components which are to be produced from forgings and the like are considered. Under these circumstances, any laboratory tests on small plain specimens should be interpreted in such a manner that the true directional properties of interest are considered.

1.7 Effect of Stress Concentration

Most designed components, of necessity, have changes in section (caused by a sharp corner, a hole, a keyway, a screwed thread, and so on), and these give rise to high localised stresses due to the effect of the consequent stress concentration. For ductile materials subjected to static loading, the effect of a *discontinuity* or *stress raiser* is usually unimportant but this is most certainly not the case when fluctuating loads are involved.

From the point of view of static considerations, it is convenient to define a *stress concentration factor, K_t* (sometimes referred to as the *theoretical* or *geometric stress concentration factor*) as the ratio of the maximum localised elastic stress to the nominal or average stress; thus, the relationship may be written as

$$K_t = \frac{\text{maximum localised elastic stress}}{\text{nominal or average stress}} \qquad (1.2)$$

The nominal or average stress can be evaluated in the usual way with the elementary equations derived from applied mechanics, but the maximum localised stress is not as easy to obtain. It can be determined for elementary geometries by mathematical means using the theory of elasticity[34], but this becomes complex for all but simple instances, and only a limited amount of theoretical data is available. Consequently, it is more usual to resort to experimental methods, typical techniques being brittle lacquer, strain gauges, photo-elasticity and photo-stress. Of these, the photoelastic method is probably still the most widely used, since it enables three-dimensional problems to be analysed[35,36].

The results obtained from investigations to determine the theoretical or geometrical stress concentration factors for a large variety of different geometries and loadings have been presented by Neuber[37] and Peterson[38].

The stress concentration factors defined by equation (1.2) are confined to static considerations where yielding does not take place. Where ductile materials are concerned, if the load is applied gradually and only a small number of times, even if local plasticity occurs at a small level, the effect of stress concentration is usually unimportant, since a redistribution of stress will occur due to the local yielding. On the other hand, for homogeneous brittle materials this is generally not true, since local yielding does not occur to any great extent, and the effect of stress concentration should be considered as being of some consequence. The usual procedure in this latter case, where sudden changes in section occur, is to multiply the nominal or average stress by the theoretical stress concentration factor in order to estimate the peak stress.

When a fluctuating load acts on a component, the effect of stress concentration can be quite severe for both ductile and brittle materials, resulting in a considerable reduction in fatigue life. As an over-simplification in the high cycle fatigue region, it is helpful to consider what happens when a ductile material having a stress raiser or notch is subjected to a fluctuating load. On the first application, some local yielding will occur at the point of stress concentration and a redistribution of stress will take place; some strain hardening or strain softening will generally accompany this redistribution (depending upon the type of material), and subsequent applications of the load will cause further yielding and cyclic strain hardening or softening. As the number of load applications is increased, it becomes more difficult for the redistribution of stress to occur and eventually a fatigue crack may develop.

In determining the reduction in fatigue strength due to the effect of stress concentration in the high cycle fatigue region, it has been common practice to define a *fatigue strength reduction factor, K_f as*

$$K_f = \frac{\text{fatigue strength of specimen without a notch}}{\text{fatigue strength of specimen with a notch}} \qquad (1.3)$$

The most obvious way of determining the value for K_f is by direct testing and a limited amount of data obtained from actual fatigue tests are available (for example, on screw threads). Apart from the fact that such tests are very time consuming and costly, the values of K_f so obtained are dependent on size and, for geometrically similar specimens of the same material, it has been established that the fatigue strength decreases with increase in size. Thus, the determination of a fatigue strength reduction factor from tests on specimens does not enable general factors for the material to be established and then to be used in design without modification.

The situation is further complicated by the fact that specimens which are geometrically identical but of different materials have varying fatigue strength reduction factors. The reason which has been suggested is that some materials are more sensitive to the presence of a notch than others, and a factor, known as the *notch sensitivity index, q,* is a measure of a materials' sensitivity to stress concentration. It may be shown[2] that the relationship between the notch sensitivity index, q, the theoretical stress concentration factor, K_t, and the fatigue strength reduction factor, K_f is given by

$$K_f = 1 + q(K_t - 1) \qquad (1.4)$$

Further, it has been proposed by Peterson[39] that

$$q = \frac{1}{1 + (a/r)} \qquad (1.5)$$

where r = notch radius, a = a material constant which depends upon grain size and tensile strength. A graphical relationship between tensile strength and the material constant a, based on test results for various steels, has been presented by Langer[40] and this is reproduced in figure 1.8.

From equation (1.5) it is seen that the notch sensitivity is a function of both material and geometry. It is interesting to observe that, as the notch radius is decreased, the geometrical stress concentration factor increases, and the notch sensitivity index decreases. Referring to equation (1.4) it is seen that since K_f is a function of both q and K_t, there must be a compensating effect taking place.

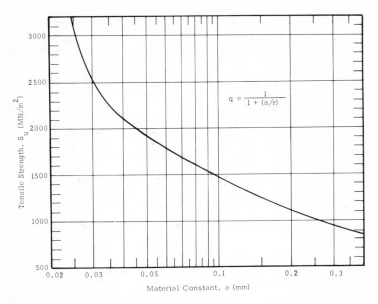

$$q = \frac{1}{1 + (a/r)}$$

Figure 1.8 *Variation of tensile strength with material constant based on investigations for steels*

It is suggested by Lipson and Juvinall[22] that the values obtained using equation (1.4) are probably too high for components not finished to the high degree produced on test specimens. Consequently, it is recommended that the fatigue strength reduction factor should be corrected for surface finish by the inclusion of the surface finish factor. Thus equation (1.4) is modfied to

$$K_f = 1 + q(K_t - 1)K_s \qquad (1.4a)$$

where K_s is the surface finish factor which can be obtained from figure 1.7.

It is pointed out that for components with surface finishes which are either as forged or hot rolled, the value of K_s to be used in equation (1.4a) should be that corresponding to a machined surface.

The effect of stress concentration at the 10^3c point is often considered to be insignificant, and in correcting the generalised fatigue curve to allow for the fatigue strength reduction factor, it is usual to divide the 10^6c point fatigue strength (i.e. the endurance limit) by K_f, to leave the fatigue strength at 10^3c unaltered, and to join these two points by a straight line on a log-log plot. There is evidence that this could lead to some error, and further discussion on the importance of stress and strain concentration in the low cycle region will be discussed in Chapter 5.

At this stage a comment about the determination of the fatigue strength reduction factor in torsion, designated K_{fs}, is required. The equations given above are applicable, but with some modification. Thus, equation (1.4a) becomes

$$K_{fs} = 1 + q_s(K_{ts} - 1)K_s \qquad (1.6)$$

where K_{ts} is obtained from the usual curves for theoretical stress concentration factors (or photomechanics), and q_s is the notch sensitivity index for torsional loading, which Peterson[39] suggests may be obtained from the equation

$$q_s = \frac{1}{1 + 0.6(a/r)} \tag{1.7}$$

where the symbols are as previously defined.

1.8 Effect of Mean Stress

From the previous discussion, one may conclude that if a reversed stress has a value less than the endurance or fatigue limit, then no matter how many times that stress is applied, failure will not occur. Unfortunately the problem is not always as simple as this. In the first instance, the mean stress is not always zero. Secondly, the stress levels obtained may be above the endurance limit, and to design so that they are reduced below the endurance limit would lead to uneconomical design. In these instances, it will be necessary to establish a working life for the component. Thus, it is evident that an interest in the complete fatigue curve with superimposed mean stresses is necessary, and attention cannot be limited to the special situations of completely reversed stresses.

The particular cases of completely reversed stresses, as obtained in the rotating bending test, are not the only ones encountered. In general, a member may be subjected to an alternating stress which varies from a maximum value σ_{max} to a minimum value σ_{min}. Such a variation with time is illustrated diagrammatically in figure 1.9. The shape of the curve will naturally depend upon the nature of the loading, and the effect of hold periods at elevated temperatures, particularly in the LCF region, will require special treatment. For the present time, only constant amplitude stress cycling at ambient conditions will be discussed. If the stress level varies from zero to a maximum value, i.e. if the minimum stress is zero, this is known as a *repeated stress*. If the stress level varies from a compressive stress to the same numerical value in tension, thus making the average stress zero, such a cycle is known as a *reversed stress*.

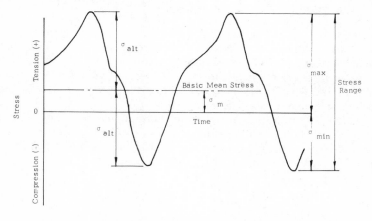

Figure 1.9 *General case of stress cycling*

Referring again to figure 1.9 it is convenient to define two new quantities; a mean or average stress σ_m, and a fluctuating or alternating stress σ_{alt}. In terms of the limiting stresses σ_{max} and σ_{min},

$$\sigma_m = \tfrac{1}{2}(\sigma_{max} + \sigma_{min}) = \frac{\sigma_{max}}{2}(1 + R) \qquad (1.8)$$

$$\sigma_{alt} = \tfrac{1}{2}(\sigma_{max} - \sigma_{min}) = \frac{\sigma_{max}}{2}(1 - R) \qquad (1.9)$$

where R is defined as the stress ratio, and is given by

$$R = \frac{\sigma_{min}}{\sigma_{max}} \qquad (1.10)$$

In the following discussion, it will be assumed that the relationship between stresses and time is sinusoidal. No account is taken of the intermediate stress levels (such as typically depicted in figure 1.9) nor of the influence of rest periods. It is further assumed that the stress cycling is periodic.

Two questions are now posed for the general case of stress cycling: what effect does the stress range have on the fatigue strength, and how is the fatigue strength affected by a mean stress? These questions were answered to a limited degree by Wöhler (1859–1870), who demonstrated that the stress range necessary to produce fracture decreases as the mean stress increases[41]. As a consequence of the work of Wöhler and others, various relationships have been proposed to account for the effect of mean stress, and in the most general form may be represented mathematically by the equation

$$\sigma_{alt} = \frac{S_e}{RF}\left[1 - \left(\frac{\sigma_m}{S_w}\right)^n\right] \qquad (1.11)$$

where S_e is the endurance limit; RF is a reserve factor, which may have the value unity under very exceptional conditions of extreme reliability; S_w is the limiting value of mean stress corresponding to zero cyclic stress; and the exponent n is a material constant. Specific proposals have been made by Gerber in 1874, by Goodman in 1899 and by Soderberg in 1930.

Gerber (Ref. 33) proposed a parabolic relationship between stress range and mean stress, in which the limiting value of mean stress corresponding to zero alternating stress was taken to be the ultimate strength of the material, S_u, expressed mathematically,

$$\sigma_{alt} = S_e\left[1 - \left(\frac{\sigma_m}{S_u}\right)^2\right] \qquad (1.12)$$

This relationship has been verified experimentally to only a limited degree.

Goodman[42] suggested a modification to Gerber's parabola which introduces additional margins of safety (the statistical nature of fatigue should be borne in mind). Expressed mathematically, the Goodman relationship is

$$\sigma_{alt} = S_e \left[1 - \frac{\sigma_m}{S_u} \right] \tag{1.13}$$

Still more popular with designers, for ductile materials, is a method proposed by Soderberg[43], in which the straight line relationship suggested by Goodman was maintained, but the limiting value for mean stress corresponding to zero alternating stress was taken to be the yield of proof stress S_p, rather than the ultimate strength. This can be expressed mathematically as

$$\sigma_{alt} = S_e \left[1 - \frac{\sigma_m}{S_p} \right] \tag{1.14}$$

Each of the three methods presented here has been used in design calculations, but probably the most popular is the Goodman line. However, where yielding is considered undesirable, the Soderberg method is commonly favoured. Although the three methods have not been verified experimentally for general application both the Goodman and the Soderberg relationships are used extensively. Soderberg's method is usually somewhat conservative, and for this reason it is often favoured. Until a more reliable method is available, considerable support exists[31] for the use of either the Goodman or the Soderberg method. The Gerber parabola is uncertain and may, in some instances, predict results which are unsafe.

A modification to the Goodman diagram which precludes both fatigue failure and yielding has been proposed by Langer[44], and this is shown in figure 1.10. The 45° line AB connects the yield strength to what Langer refers to as the limit of elastic behaviour, which is described as the highest stress amplitude that the material can withstand without yielding, even after cyclic loading. For all practical

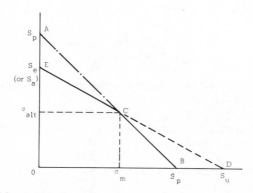

Figure 1.10 *Langer's modification to the Goodman diagram in the LCF region*

purposes this may be taken as the 0.2 per cent offset yield strength (designated S_p in the diagram). Thus, according to the Langer modification, any combination of mean and alternating stress must lie within that part of the diagram enclosed by OBCE.

So far, it is evident that the effect of a mean stress is to decrease the allowable alternating stress which may be applied without failure. There is an exception to this, however, which must be noted. This exception is for compressive mean stresses which, provided they do not exceed the yield point in compression, do not diminish the allowable alternating stress. In fact, there is some evidence to suggest that the effect of a compressive mean stress is to increase fatigue resistance. For purposes of design calculations, provided that the compressive mean stress does not exceed the yield point in compression, it is suggested that the fatigue strength be assumed the same as if zero mean stress were acting.

As a final point, it should be mentioned that the use of general reserve factors obtained using statistical data[2] can readily be accommodated in the analysis simply by dividing the fatigue strength and the limiting mean stress (i.e. the ultimate or yield strength, as appropriate) by the values for RF. Thus, suppose that the reserve factor desirable for the alternating stress is RF_a, and for the mean stress RF_m, then using the Goodman relationship represented by equation (1.13) we have

$$\sigma_{alt} = \frac{S_a}{RF_a} \left[1 - \frac{\sigma_m}{S_u/RF_m} \right] \qquad (1.15)$$

There is no reason why the reserve factors applied to the mean and alternating stresses should have the same numerical values. In fact, it would seem reasonable that the value of RF_a should be somewhat greater than that of RF_m, because of the inherent scatter and consequential greater unreliability associated with fatigue data. However, much depends upon the reliability of the component fatigue curve, and each case must be judged on its merits.

For convenience, assume that $RF_a = RF_m = RF$, and transpose equation (1.15), so that in this case,

$$RF = \frac{S_u}{\sigma_m + (S_u/S_a)\sigma_{alt}} \qquad (1.16)$$

$$= \frac{S_u}{\sigma_m + c\sigma_{alt}} \qquad (1.17)$$

where $c = S_u/S_a$, and is a combined material and geometric constant for a particular region in a real component. Note that the value for RF may be assessed using the philosophy of confidence levels, and if the calculated stresses and the material properties are known with a confidence level of 100 per cent (not possible in reality), the value for RF would be unity. As stated previously, the actual value to be used can be assessed using statistical data, and typically a value in the region of about 1.3–2 is reasonable. Values in excess of this range are frequently used in practice, but it must be recognised that in such instances the design is either over-conservative, or else the analysis is at the very best only approximate.

The factor of safety, or reserve factor, is defined on a static basis according to whether fracture or yielding is the limiting criterion. Thus the reserve factor is then expressed as

$$RF = \frac{S_u}{\sigma_s} \qquad (1.18)$$

or

$$RF = \frac{S_p}{\sigma_s} \qquad (1.19)$$

depending upon whether the ultimate strength or the yield strength is appropriate. In equations (1.18) and (1.19), σ_s is the maximum static stress acting. Now by analogy, for the same risk factor under static and dynamic loading, it is seen that the denominator in equation (1.17) may be considered as an equivalent static load stress having the same damaging effect as the actual defined fluctuating load system. In effect, this means that for a fluctuating load system, if the mean and alternating stress components are replaced by an equivalent static stress having the same damaging effect, then the subsequent analysis may be performed on a static basis. This approach has no advantages when we are considering only a one-dimensional stress system, but it is extremely useful for dealing with combined stresses, as will be demonstrated presently. The equivalent static stress having the same damaging effect as a mean and alternating combination is expressed as

$$\sigma_s = \sigma_m + c\sigma_{alt} \qquad (1.20)$$

One final point to mention, is that although the discussions in this section have been presented in terms of direct stresses, the expressions are equally applicable for torsional or shear stresses, but of course, the appropriate material properties must be used.

1.9 Environmental Factors

Environmental factors, more particularly that of temperature, may have a profound effect on the fatigue resistance of a component, since material behaviour varies widely with operating temperature. For temperatures in excess of about 0.35—0.7 times the absolute melting point of the base metal, progressive deformation or creep becomes important and it is likely that in these instances the criterion of failure will be that of deformation. Where cyclic loading occurs at such temperatures, creep—fatigue interactions will be involved and the effect of strain rate, frequency and dwell periods will be important.

For operating conditions at elevated temperature below the creep range, the methods previously outlined in the present chapter are applicable in the HCF region. However, the effect of elevated temperature is to reduce the fatigue strength in the HCF region, and in assessing the fatigue integrity of a component it is of obvious importance to use material properties corresponding to the temperature of interest.

If the temperature is sufficiently high to make creep a significant factor, it has been suggested that a conservative approximation to material or component behaviour may be obtained using a modified version of the Goodman or Gerber method outlined in Section 1.8. The fatigue strength at the temperature of interest is plotted as the ordinate and the limiting mean stress is taken as the creep rupture strength, as shown in figure 1.11. Further details will be presented in a later chapter and it is suggested that better correlation between predicted and experimental values is obtained using the equation of an ellipse.

The effect of low operating temperatures on a material is generally to increase the ultimate and yield strengths, but this is usually accompanied by a reduction in ductility, and there is a tendency for failure to occur by so-called *brittle fracture*. For most ferrous metals, there is a sudden transition from ductile to brittle fracture as the temperature is reduced, the temperature at which this change occurs being known as the *ductile—brittle transition*.

Over the last ten or fifteen years fracture mechanics has been studied and

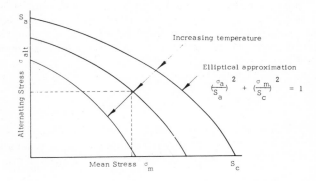

Figure 1.11 *Effect of mean stress for combined creep and fatigue*

developed as a means of establishing the critical crack length at which fast fracture occurs, and not only is this critical length important, but the rate of crack propagation is of great practical significance; this latter statement is nearer the truth for low-cycle failure conditions. A more detailed discussion concerned with fracture mechanics and its application to problems of fatigue will be presented later, but at the present time it is interesting to mention the effect of temperature on crack growth. This is indicated schematically in figure 1.12. It is seen that as the temperature is reduced, the proportion of life spent in initiation is increased, the number of cycles required to cause failure, in general, increases but the critical length at which fast fracture occurs decreases.

The deterioration of a material due to corrosion has a very significant detrimental effect on the material's strength. When combined with cyclic loading, failure is likely to be accelerated, and the reduction in life will probably be considerably greater than would be expected from an estimate of the two factors acting separately. Consequently, corrosion fatigue is just one example where the principle of superposition does not apply. However, under such conditions of combined corrosion fatigue, the basic resistance of the material to this combined environment is largely governed by its resistance to corrosion. That is to say, the effect of increasing the material's resistance to corrosion is likely to be more

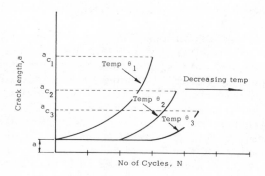

Figure 1.12 *Crack growth plotted against number of stress cycles*

significant than increasing its fatigue strength, although improving both is obviously desirable.

A particular form of corrosion fatigue may occur in situations which involve the relative movement of contacting surfaces under the action of an alternating load; this is known as fretting corrosion, and may occur in all types of joints or fitted assemblies. It may be combatted by several methods, such as the use of anti-fretting compounds; elimination of the relative motion; reduction in the surface tensile stresses; surface hardening or separation of adjacent parts. Of course, certain combinations of materials are more susceptible to fretting corrosion than others, and material compatibility to this form of attack is most important. Comments relating to preferred combinations and information generally on specific materials, has been presented by Heywood[21]. For more comprehensive details relating to mechanisms of corrosion fatigue and fretting corrosion, the reader is referred to the work of Forrest[46], Cazaud[15] and Nicholson[47] (see also chapter 4).

For applications in the field of nuclear engineering, such as the components of nuclear reactors and the like, where materials are subjected to intense radioactive radiation and neutron bombardment, the effect of such environment must be taken into account. In general, for most constructional materials, the effect of radiation is to increase hardness and tensile strength and reduce ductility, thus making the material more rigid and brittle, and consequently more likely to fail by brittle fracture. In addition, an increase in the creep rate occurs, and the allowable fatigue stress corresponding to a particular life is reduced[48]. The significance of the changes depend upon the material and the other environmental effects of temperature and flux intensities, and for many alloys certainly requires consideration[49].

1.10 Concluding Remarks

The discussions undertaken in this chapter have, in the main, been concerned only with one-dimensional stress situations, and this, of course, is a severe limitation when the application to real situations is undertaken. However, the extension to service components subjected to dynamic loading involving complex stresses will be presented in another chapter, when it will be evident that the discussions presented here are still valid.

Further complications arise when the loads are not of constant amplitude, and determining the anticipated loading history is often just as complex as assessing the fatigue integrity once this history is known. In many cases the only reliable way of estimating load history is by actually measuring the loads under service conditions. With the advent of servo-controlled fatigue testing machines, this anticipated loading history, if known or predicted, may be accurately reproduced on either a test specimen or a component itself. Although considerable research into the effect of cumulative damage has been carried out[48] and particular relationships have been developed for special conditions and materials[50,51], the Palmgren—Miner relationship[52,53] still remains the basis for assessing fatigue damage under non-constant amplitude loading situations, at the design stage. In equation form, this may be written as

$$UF = \sum_{i=1}^{L} \frac{N}{N_f} \tag{1.21}$$

Strictly speaking, according to the Palmgren—Miner relationship, the value for UF is unity, and it is assumed that damage accumulates linearly. In equation (1.21), N is the number of loading applications at a particular stress or strain level. In practice, it has been found that damage does not always accumulate linearly, and that the loading sequence can often have a very significant effect on the way in which damage actually accumulates. This is one of the main criticisms of equation (1.21). Nevertheless, from a design point of view it still has a great deal of attraction because of its simplicity. In reality, the usage factor must be recognised as being dependent upon sequence of loading and the actual material, and for most practical situations it is commonly assumed that UF should not exceed about 0.8, although for conservative applications, a value as low as 0.3 has even been suggested[50].

Other factors which affect fatigue behaviour will be discussed in other chapters. The application will be demonstrated in connection with the design of some real components.

References

1. Duggan, T. V., Current Trends in Fatigue Research, *The Chartered Mechanical Engineer*, **17** (10), London (1970)
2. Duggan, T. V., *Applied Engineering Design and Analysis*, Iliffe, London (1970)
3. Duggan, T. V., Fatigue Assessment of a Simulated Component, *Tech. Report No. F.303*, Portsmouth Polytechnic, Portsmouth (1970)

4. Peterson, R. E., Fatigue of Metals – Part 3 – Engineering Design Aspects, *Matls. Res. and Standards*, **3**, 122 (1963)
5. Manson, S. S., *Thermal Stress and Low Cycle Fatigue*, McGraw-Hill, New York (1966)
6. Yokobori, T., *Strength, Fracture and Fatigue of Materials*, Noordhoff, Groningen, The Netherlands (1964)
7. Duggan, T. V., Lowcock, M. T. and Spence, L. J., An Investigation into the Fatigue Strength of a Creep Resisting Stainless Steel Under Constant Amplitude Rotating Bending Loads, *Tech. Mem. PCT/Mech. Eng./RR1*, Portsmouth Polytechnic, Portsmouth (1969)
8. Cicci, F., An Investigation of the Statistical Distribution of Constant Amplitude Fatigue Endurances for a Maraging Steel, *Utias Tech. Note No. 73*, Inst. for Aerospace Studies, University of Toronto (1964)
9. Freudenthal, A. M. and Grumble, E. J., Distribution Functions for the Prediction of Fatigue Life and Fatigue Strength, *Intern. Conf. on Fatigue*, Inst. Mech. Engrs, London (1956)
10. Weibull, W., *Fatigue Testing and Analysis of Results*, Pergamon Press, New York (1961)
11. Weibull, W., Analysis of Fatigue Test Results, *First Seminar on Fatigue Design*, Columbia University, New York (1963)
12. Finney, J. M., A Review of the Discontinuity or Hump Phenomenon in Fatigue S/N Curves: Theories and Further Results, *Structures and Materials Report 314*, Australian Defence Scientific Service, Melbourne (1967)
13. Juvinall, R. C., *Engineering Consideration of Stress, Strain and Strength*, McGraw-Hill, New York (1967)
14. Esin, A., The Microplastic Strain Energy Criterion Applied to Fatigue (Ph.D. Dissertation), University College, London (1966)
15. Cazaud, R., *Fatigue of Metals* (Trans. Fenner, A. J.), Chapman and Hall, London (1953)
16. Dolan, T. J., Stress Range, *Metals Engineering – Design* (A.S.M.E. Handbook), Pt. 2, Section 6.2, McGraw-Hill, New York (1952)
17. Morrison, J. M., The Yield of Mild Steel with Particular Reference to the Effect of Size of Specimen, *Proc. Instn Mech. Engrs*, **142**, 193 (1939)
18. Morkovin, D. and Moore, H. F., Third Progress Report on the Effect of Size of Specimen on Fatigue Strength of Three Types of Steel, *Proc. Am. Soc. Test. Matls*, **44**, 137 (1944)
19. Phillips, C. E. and Heywood, R. B., The Size Effect of Plain and Notched Steel Specimens Loaded Under Reversed Direct Stress, *Proc. Instn Mech. Engrs*, **165** (W.E.P.65), 113 (1951)
20. Pope, J. A., *Metal Fatigue*, Chapman and Hall, London (1959)
21. Heywood, R. B., *Designing Against Fatigue,* Chapman and Hall, London (1962)
22. Lipson, C. and Juvinall, R. C., *Handbook of Stress and Strength*, Macmillan, New York (1963)
23. Coyle, M. B. and Watson, S. J., Fatigue Strength of Turbine Shafts with Shrunk-on Discs, *Proc. Instn Mech. Engrs*, **178**, 147–183 (1964)
24. Lipson, C., Noll, G. C. and Clock, L. S., Significant Strength of Steels in the Design of Machine Parts, *Product Engineering*, **20** (4), 142 (5), 124 (1949)

25. Hanley, B. C. and Dolan, T. J., Surface Finish, *Metals Engineering – Design* (A.S.M.E. Handbook), Pt. 2, Section 6.5, McGraw-Hill, New York (1953)
26. Sines, G. and Waismann, J. L. (Eds.), *Metal Fatigue*, McGraw-Hill, New York (1959)
27. Mattson, R. L. and Roberts, J. G., Effect of Residual Stress Induced by Shot Peening upon Fatigue Strength, *Internal Stresses and Fatigue in Metals* (ed. Rassweiler, G. M. and Grube, W. L.), Elsevier, New York (1959)
28. Almen, J. O. and Black, P. H., *Residual Stresses and Fatigue in Metals*, McGraw-Hill, New York (1963)
29. Faires, V. M., *Design of Machine Elements*, 4th edn, Macmillan, New York (1965)
30. Horger, O. J., Cold Working, *Metals Engineering – Design*, (A.S.M.E. Handbook) Pt. 2, Section 6.9, McGraw-Hill, New York (1953)
31. Grover, H., Gordon, S. A. and Jackson, L. R., *Fatigue of Metals and Structures* (NAVWEPS 00-25-534), U.S. Government Printing Office, Washington (1960)
32. Frith, P. H., Fatigue Tests on Crankshaft Steels, Part 1: The Effect of Nitriding on the Fatigue Properties of Chromium—Molybdenum Steel, *J. Iron and Steel Inst.*, **159**, 385 (1948)
33. Lessels, J. M., *Strength and Resistance of Metals*, Wiley, New York (1954)
34. Timoshenko, S. P. and Goodier, J. N., *Theory of Elasticity*, 2nd edn, McGraw-Hill, New York (1951)
35. Hetenyi, M. (Ed.), *Handbook of Experimental Stress Analysis*, Wiley, New York (1950)
36. Heywood, R. B., *Photoelasticity for Designers*, Pergamon Press, Oxford (1969)
37. Neuber, H., *Theory of Notch Stresses, Principles and Exact Calculations* (English Translation), Edwards, Ann Arbor, Michigan (1946)
38. Peterson, R. E., *Stress Concentration Design Factors*, Wiley, New York (1953)
39. Peterson, R. E., Notch Sensitivity, Chapter 13, *Metal Fatigue*, (ed. Sines, G. and Waismann, J. L.), McGraw-Hill, New York (1959)
40. Langer, B. F., Application of Stress Concentration Factors, *Bettis Tech. Rev. WAPD – BT – 18*, 1 (1960)
41. Wöhler, A., Uber die festigkeitsverusche mit eisen und stahl, *Z. fur Bauwesen*, **8** (1858), **10, 13, 16** and **20** (1870) (Tests to Determine the Forces Acting on Railway Carriage Axles and the Capacity of Resistance of the Axles, English Abstract, *Engineering*, **11**, 199 (1971))
42. Goodman, J., *Mechanics Applied to Engineering*, vol. 1, 9th edn, p. 634, Longmans, London (1930)
43. Soderberg, C. R., Factor of Safety and Working Stress, *Trans. Am. Soc. Test Matls*, **52** (2) (1930); Working Stresses, *Trans. Am. Soc. Test. Matls*, **57**, A 106 (1935)
44. Langer, B. F., Design of Pressure Vessels for Low Cycle Fatigue, *J. Basic Engng*, **84** (3), 389 (1962)
45. Smith, J. O., The Effect of Range of Stress on the Fatigue Strength of Metals, *Bull. Ill. Univ. Engng Exp. Stn*, **316** (37), 5 (1939) and **334** (39), 26 (1942)
46. Forrest, P. G., *Fatigue of Metals*, Pergamon Press, Oxford (1962)
47. Nicholson, C. E., Influence of Mean Stress and Environment on Crack Growth, *Mechanics and Mechanisms of Crack Growth*, Cambridge University, 4—6th April (1973)

48. Osgood, C. C., *Fatigue Design*, Wiley-Interscience, New York (1970)
49. Ilyushin, A. A. and Lensky, V. S., *Strength of Materials*, Pergamon Press, Oxford (1967)
50. Madayag, A. F. (Ed.), *Metal Fatigue: Theory and Design*, Wiley, New York (1969)
51. Brook, R. H. W., An Investigation of Cumulative Fatigue Damage Using a Non-Destructive Measurement of Fatigue Lives (*Ph.D. Dissertation*), University of Bristol (1967)
52. Palmgren, A., Life of Ball Bearings, *Z.V.D.I.*, **68**, 339 (1924)
53. Miner, M. A., Cumulative Damage in Fatigue, *J. appl. Mechanics*, **12** (3), A159 (1945)

2

Fatigue Analysis of Combined Stress Systems

2.1 Introduction

In chapter 1 the factors affecting fatigue behaviour were discussed, and whilst many of the comments made were quite general in their applicability, the fatigue analysis of combined stress systems was not considered. In particular, the combination of mean and alternating stresses was dealt with only for one-dimensional stress systems, and the usual equations from elementary applied mechanics enable such stresses to be calculated. With the aid of a modified fatigue curve for the component of interest, fatigue integrity can readily be assessed.

Under the action of combined dynamic stresses, the situation is somewhat more complicated, and it is necessary to establish methods of analysis for assessing the fatigue integrity of components when combined stresses are involved. From a fundamental standpoint, there is no reason to suppose that the same theories which govern static failure are applicable under dynamic loading. Nevertheless, it would seem appropriate to attempt an extension of the usual classical theories to include fatigue, and this is the basis for the methods proposed in this chapter. In the main, the discussions will be concerned with the HCF region, including some fatigue—creep interaction considerations, and the analysis for the LCF region will be undertaken in a subsequent chapter.

2.2 General Case of Three-Dimensional Stresses

Consider the element shown in figure 2.1, subjected to direct stresses σ_x, σ_y and σ_z, and shear stress τ_{xy}, τ_{xz} and τ_{yz}. Before proceeding with the analysis for such a system it is desirable to establish and clarify the notation. Consider some point O in a material, and through this point define rectangular co-ordinates in the x, y and z directions, as indicated in figure 2.2.

Now consider an elemental area δA in the x-y plane. In general, there will be a resultant force δP acting across the element, which can be resolved into components δP_x, δP_y and δP_z acting in the x, y and z directions respectively. If the elemental area is sufficiently small, the force δP may be regarded as uniformly distributed. Thus, in the limit

$$\frac{\delta P_x}{\delta A} = \tau_{zx}; \qquad \frac{\delta P_y}{\delta A} = \tau_{zy}; \qquad \frac{\delta P_z}{\delta A} = \sigma_z$$

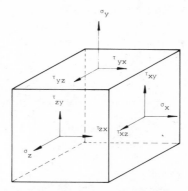

Figure 2.1 *Element under simplified three-dimensional stress*

where the symbols τ and σ have been used to represent shear stresses (i.e. stresses parallel to the plane) and normal stresses (i.e. stresses perpendicular to the plane) in the usual way. The double suffix for shear stress is used to indicate the direction of the shear stress and the plane on which it acts. Thus, τ_{zx} indicates a shear stress on the plane perpendicular to the z axis (i.e. the x-y plane) and acting in the x direction, whilst τ_{zy} indicates a shear stress on the plane perpendicular to the z axis, but acting in the y direction. The suffix used for the normal stress component

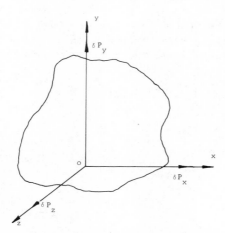

Figure 2.2 *Co-ordinate system for three-dimensional stress*

simply indicates the direction of the stress. Similarly, then, σ_x and σ_y define normal stresses in the x and y directions respectively; τ_{xy} and τ_{yz} define shear stresses in the plane perpendicular to the x axis (i.e. the y-z plane) in the directions y and z respectively; and τ_{yz} and τ_{zy} define shear stresses in the plane perpendicular to the y axis (i.e. the x-z plane) in the directions x and z respectively. The above discussion should make quite clear the notation indicated in figure 2.1.

Thus, the state of stress is defined by the nine components

$$\begin{pmatrix} \sigma_x & \tau_{xy} & \tau_{xz} \\ \tau_{yx} & \sigma_y & \tau_{yz} \\ \tau_{zx} & \tau_{zy} & \sigma_z \end{pmatrix}$$

which is termed the *stress tensor*. However, it is easily shown[1,2] that

$$\begin{aligned} \tau_{xy} &= \tau_{yx} \\ \tau_{xz} &= \tau_{zx} \\ \tau_{yz} &= \tau_{zy} \end{aligned}$$

making the stress tensor symmetrical.

Now consider a plane ABC oblique to the axes, as shown in figure 2.3, such that ABCO defines a small tetrahedron with three edges along the co-ordinate axes, and the plane ABC having its normal defined by direction cosines l, m and n. To make this clear let the normal to ABC passing through the point O be located at a point N on the surface of ABC, so that ON defines the normal to ABC drawn through the point O. Thus, the direction cosines are defined as

$$l = \frac{ON}{OA}; \qquad m = \frac{ON}{OB}; \qquad n = \frac{ON}{OC}$$

Let the area ABC be denoted by δA and the areas OBC, OAC and OAB by δA_x, δA_y and δA_z respectively. It is convenient to represent these latter three elemental areas in terms of the area δA, and this can be done using simple geometry. Thus,

$$\begin{aligned} \text{Volume of ABCO} &= \delta A \ \times \frac{ON}{3} \\ &= \delta A_x \times \frac{OA}{3} \\ &= \delta A_y \times \frac{OB}{3} \\ &= \delta A_z \times \frac{OC}{3} \end{aligned}$$

from which

$$\delta A_x = \frac{ON}{AO} A = l\delta A$$

$$\delta A_y = \frac{ON}{OB} A = m\delta A$$

$$\delta A_z = \frac{ON}{OB} A = n\delta A$$

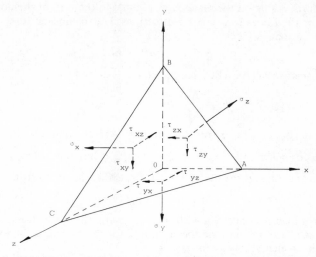

Figure 2.3 *Tetrahedron defining an oblique plane for three-dimensional stresses*

Resolving forces in the x, y and z directions (but not including forces acting in the plane ABC) and designating these forces P_x, P_y and P_z respectively, the following results are easily obtained:

$$P_x = \sigma_x \delta A_x + \tau_{yx} \delta A_y + \tau_{zx} \delta A_z$$
$$= \delta A(l\sigma_x + m\tau_{yx} + n\tau_{zx}) = \delta A_{p_x} \tag{2.1a}$$

$$P_y = \sigma_y \delta A_y + \tau_{xy} \delta A_x + \tau_{zy} \delta A_z$$
$$= \delta A(m\sigma_y + l\tau_{xy} + n\tau_{zy}) = \delta A_{p_y} \tag{2.1b}$$

$$P_z = \sigma_z \delta A_z + \tau_{xz} \delta A_x + \tau_{yz} \delta A_y$$
$$= \delta A(n\sigma_z + l\tau_{xz} + m\tau_{yz}) = \delta A_{p_z} \tag{2.1c}$$

Consequently, the resultant force on ABC for equilibrium (say P) is

$$P = [P_x^2 + P_y^2 + P_z^2]^{1/2} \tag{2.2}$$

and the so-called *resultant stress* on ABC (say p) is

$$p = \frac{p}{\delta A} = [p_x^2 + p_y^2 + p_z^2]^{1/2} \tag{2.3}$$

where

$$p_x = l\sigma_x + m\tau_{yx} + n\tau_{zx}$$
$$p_y = m\sigma_y + l\tau_{xy} + n\tau_{zy} \tag{2.4}$$
$$p_z = n\sigma_z + l\tau_{xz} + m\tau_{yz}$$

The normal force acting on the plane ABC is equal to the sum of the normal components of the forces in x, y and z directions. Let the normal force be designated N, so that

$$N = lP_x + mP_y + nP_z$$

The normal stress acting on ABC, designated σ_n, is given by

$$\sigma_n = \frac{N}{\delta A} = lp_x + mp_y + np_z \tag{2.5}$$

Substituting for p_x, p_y and p_z from equations (2.4) into equation (2.5) gives

$$\sigma_n = l^2 \sigma_x + m^2 \sigma_y + n^2 \sigma_z + 2lm\tau_{xy} + 2ln\tau_{xz} + 2mn\tau_{yz} \tag{2.6}$$

The shear stress acting on the plane ABC, say τ, is then obtained from

$$\tau = [p^2 - \sigma_n^2]^{1/2} \tag{2.7}$$

The designer is mainly concerned with the maximum values for stress, and equation (2.5) can be studied to determine the condition for the normal stress to be a maximum. From elementary geometry[3],

$$l^2 + m^2 + n^2 = 1 \tag{2.8}$$

If l and m are treated as independent variables, then for a maximum or minimum value

$$\frac{\partial \sigma_n}{\partial l} = 0; \qquad \frac{\partial \sigma_n}{\partial m} = 0$$

Therefore, from equation (2.6)

$$\frac{\partial \sigma_n}{\partial l} = 2l\sigma_x + 2n \frac{\partial n}{\partial l} \sigma_z + 2m\tau_{xy} + 2n\tau_{xz} + 2l \frac{\partial n}{\partial l} \tau_{xz} + 2m \frac{\partial n}{\partial l} \tau_{yz} = 0$$

therefore,

$$p_x + \frac{\partial n}{\partial l} p_z = 0 \tag{2.9a}$$

$$\frac{\partial \sigma_n}{\partial m} = 2m\sigma_y + 2n \frac{\partial n}{\partial m} \sigma_z + 2l \frac{\partial n}{\partial m} \tau_{xz} + 2n\tau_{yz} + 2l\tau_{xy} + 2m \frac{\partial n}{\partial m} \tau_{yz} = 0$$

therefore,

$$p_y + \frac{\partial n}{\partial m} p_z = 0 \tag{2.9b}$$

Differentiating equation (2.8) with respect to l gives

$$\frac{\partial n}{\partial l} = -\frac{l}{m} \tag{2.10a}$$

and differentiating equation (2.8) with respect to m gives

$$\frac{\partial n}{\partial m} = -\frac{m}{n} \tag{2.10b}$$

Substituting from equations (2.10) into equations (2.9) gives

$$\frac{P_x}{l} = \frac{P_y}{m} = \frac{P_z}{n} \tag{2.11}$$

Substituting for p_y and p_z in terms of P_x from equation (2.11) into equation (2.3) for resultant stress gives

$$p_x = lp$$
$$p_y = mp \tag{2.12}$$
$$p_z = np$$

Consequently, the resultant stress is in the direction (l, m, n) i.e. normal to the plane, indicating that there is no shear stress acting on the plane; such a plane is a principal plane and the normal stress acting is a principal stress. Using the notation σ_n to replace p, indicating a normal stress, and substituting for p_x, p_y and p_z from equations (2.4) into equation (2.12)

$$\sigma_n l = l\sigma_x + m\tau_{yx} + n\tau_{zx}$$
$$\sigma_n m = m\sigma_y + l\tau_{xy} + n\tau_{zy}$$
$$\sigma_n n = n\sigma_z + l\tau_{xy} + m\tau_{yz}$$

Rearranging and noting that $\tau_{yx} = \tau_{xy}, \tau_{zx} = \tau_{xz}$, and $\tau_{zy} = \tau_{yz}$

$$(\sigma_n - \sigma_x)l - \tau_{xy}m - \tau_{xz}n = 0$$
$$-\tau_{xy}l + (\sigma_n - \sigma_y)m - \tau_{yz}n = 0 \tag{2.13}$$
$$-\tau_{xz}l - \tau_{yz}m + (\sigma_n - \sigma_z)n = 0$$

These equations may be most elegantly expressed in determinant form, and for a non-trivial solution

$$\begin{vmatrix} (\sigma_n - \sigma_x) & -\tau_{xy} & -\tau_{xz} \\ -\tau_{xy} & (\sigma_n - \sigma_y) & -\tau_{yz} \\ -\tau_{xz} & -\tau_{yz} & (\sigma_n - \sigma_z) \end{vmatrix} = 0 \tag{2.14}$$

Solving this determinant in the usual way[4] and putting it equal to zero yields the following cubic equation in σ_n:

$$\sigma_n^3 - (\sigma_x + \sigma_y + \sigma_z)\sigma_n^2 - (\tau_{xy}^2 + \tau_{yz}^2 + \tau_{xz}^2 - \sigma_x\sigma_y - \sigma_y\sigma_z - \sigma_x\sigma_z)\sigma_n$$
$$- (\sigma_x\sigma_y\sigma_z + 2\tau_{xy}\tau_{yz}\tau_{xz} - \sigma_x\tau_{yz}^2 - \sigma_y\tau_{xz}^2 - \sigma_z\tau_{xy}^2) = 0$$

It is convenient to write this equation in the form

$$\sigma_n^3 - I_1\sigma_n^2 - I_2\sigma_n - I_3 = 0 \tag{2.15}$$

where

$$I_1 = (\sigma_x + \sigma_y + \sigma_z) \tag{2.16}$$
$$I_2 = (\tau_{xy}^2 + \tau_{yz}^2 + \tau_{xz}^2 - \sigma_x\sigma_y - \sigma_y\sigma_z - \sigma_x\sigma_z) \tag{2.17}$$
$$I_3 = (\sigma_x\sigma_y\sigma_z + 2\tau_{xy}\tau_{yz}\tau_{xz} - \sigma_x\tau_{yz}^2 - \sigma_y\tau_{xz}^2 - \sigma_z\tau_{xy}^2) \tag{2.18}$$

and are termed the *invariants*, since they obviously do not change with co-ordinate axes, i.e. the principal stresses will remain the same for the same stress system.

In the special case of a general co-planar stress system $\sigma_z = 0$, $\tau_{yz} = \tau_{xz} = 0$, and in this instance

$$I_1 = (\sigma_x + \sigma_y)$$
$$I_2 = (\tau_{xy}^2 - \sigma_x \sigma_y)$$
$$I_3 = 0$$

Substituting these values into equation (2.15), we obtain

$$\sigma_n^2 - (\sigma_x + \sigma_y)\sigma_n - (\tau_{xy}^2 - \sigma_x \sigma_y) = 0$$

and solving by the usual quadratic equation

$$\sigma_n = \tfrac{1}{2}(\sigma_x + \sigma_y) \pm \tfrac{1}{2}[(\sigma_x - \sigma_y) + 4\tau_{xy}^2]^{1/2} \tag{2.19}$$

Equation (2.15) has three real roots, the values of which are the three principal stresses σ_1, σ_2 and σ_3. This equation may be solved by the usual methods of trial and error or graphical methods, or by suitable method of approximations, such as the Newton–Raphson method[4]. This latter method may be conveniently represented by the equation

$$\sigma_{(r+1)} = \sigma_r - \frac{f(\sigma_r)}{f'(\sigma_r)} \tag{2.20}$$

where σ_r = first approximation (guessed or obtained graphically),
        $f(\sigma_r)$ = original function with assumed value σ_r,
        $f'(\sigma_r)$ = first differential of the function with σ_r equal to the assumed value.

An iterative procedure is carried out until the correct values are obtained. Equation (2.20) may be expressed, for this particular application, in the form

$$\sigma_{(r+1)} = \sigma_r - \frac{\sigma_r^3 - I_1\sigma_r^2 - I_2\sigma_r - I_3}{3\sigma_r^2 - 2I_1\sigma_r - I_2} \tag{2.21}$$

Thus, the analysis indicates that any three-dimensional stress system can be reduced to that of three principal stresses acting on mutually perpendicular planes, and the strain in each direction can be calculated[1] from the results.

$$\epsilon_1 = \frac{1}{E}[\sigma_1 - \nu(\sigma_2 + \sigma_3)]$$

$$\epsilon_2 = \frac{1}{E}[\sigma_2 - \nu(\sigma_1 + \sigma_3)] \tag{2.22}$$

$$\epsilon_3 = \frac{1}{E}[\sigma_3 - \nu(\sigma_1 + \sigma_2)]$$

The maximum shear stress values can be calculated from the differences of principal stresses, i.e.

$$\tau_{12} = \pm \frac{\sigma_1 - \sigma_2}{2}$$

$$\tau_{23} = \pm \frac{\sigma_2 - \sigma_3}{2} \qquad (2.23)$$

$$\tau_{13} = \pm \frac{\sigma_1 - \sigma_3}{2}$$

2.3 Static Theories of Failure

When an elastic body is subjected to a system of combined stresses it may fail according to one of a number of different limiting conditions. There are diverse opinions on the physical cause of failure, and many different theories have been proposed. Some of these proposals have been shown to give results divergent from those obtained in practice.

Any discussion relating to failure demands first a definition of what is meant by failure, and then a consideration of the means by which it comes about. In a general sense, failure may be defined as a condition whereby the component either becomes unsafe or can no longer perform its required function in a satisfactory manner. This may correspond to excessive yielding or deformation, deterioration due to wear, the presence of a crack of a certain size, or by complete fracture. In the present discussion, attention is limited to combined stress systems subjected to static loading, and it is only necessary to point out that failure and fracture may be two distinct phenomena.

In the so-called *classical theories of failure*, attempts are made to predict combined stress behaviour on the basis of data obtained from simple tensile or compressive tests. In converting the unidirectional data to combined stress data, five basic theories of failure have been considered, each one in its original form assuming that failure is deemed to occur when an elastic condition is exceeded. The mechanism of failure is complex and the particular manner in which an elastic condition may be exceeded depends upon the stress system and the material behaviour.

A material subjected to an axial tensile stress experiences the following phenomena:

(1) direct stress on planes normal to the direction of the applied stress;
(2) direct and shear stresses on planes inclined to the direction of the applied stress;
(3) as a consequence of (1) and (2), direct and shear strains; and
(4) energy is stored in the material.

When the material reaches the elastic limit (or some other limiting quantity), the above quantities all have particular values, and the problem is not only to calculate these values, but to determine for particular materials which of these quantities, or combinations, cause the elastic limit (or some other critical value) to be exceeded. It is this question which the classical failure theories attempt to answer when considering a system subjected to combined stresses. Basic assumptions made in all the theories are that the material remains homogeneous and isotropic.

2.3.1 Maximum Principal Stress Theory

This theory is due to Rankine and assumes that failure will occur when the maximum principal stress is equal to a certain limiting quantity or critical value, S_c. Under static loading conditions this is taken as the stress corresponding to the elastic limit, more particularly the yield or proof stress, S_p. Thus, according to this theory, failure is deemed to occur when

$$\sigma_1 \geqslant S_c \tag{2.24}$$

where, in the special case, $S_c = S_p$.

Based on experimental data obtained for cast irons and other brittle materials, it is commonly assumed the maximum principal stress theory is applicable for brittle materials, provided that the values for principal stresses do not have different signs. However, it has been suggested[5], that recent data shows that brittle material can be handled by the distortion energy theory when allowance is made for the stress concentrating effects of inclusions which often give rise to lack of ductility in a part.

If σ_1 is the criterion of failure, substituting from equation (2.19) into equation (2.24) and transposing gives the result

$$\left(\frac{\sigma_x + \sigma_y}{S_c}\right) + \left(\frac{\tau_{xy}}{S_c}\right)^2 - \left(\frac{\sigma_x \sigma_y}{S_c^2}\right) = 1 \tag{2.25}$$

and in the special case when $\sigma_y = 0$, this reduces to

$$\left(\frac{\sigma_x}{S_c}\right) + \left(\frac{\tau_{xy}}{S_c}\right)^2 = 1 \tag{2.26}$$

2.3.2 Maximum Principal Strain Theory

According to this theory, suggested by St. Venant, failure is deemed to occur when the maximum principal strain of the complex system is equal to the maximum strain corresponding to the elastic limit (or some other critical value). Stated mathematically, failure will occur when (see equations (2.22))

$$\epsilon_c = \frac{S_c}{E} = \frac{1}{E} \left[\sigma_1 - \nu(\sigma_2 + \sigma_3)\right]$$

which reduces to

$$S_c = \sigma_1 - \nu(\sigma_2 + \sigma_3) \tag{2.27}$$

where, in the special case $S_c = S_p$.

Results obtained by this theory are not always valid, and it is not now in general use.

2.3.3 Maximum Shear Stress Theory

This theory was first suggested by Coulomb but it is usually attributed to Guest. It assumes that failure occurs when the maximum shear stress for a complex stress system is equal to a certain critical value. Under static loading, this critical value is usually taken as being equal to the shear stress at the elastic limit in simple tension. From test results, the approximation that the shear stress in simple tension corresponding to yield is equal to $S_p/2$ is justified for most applications. Thus, by this theory failure is deemed to occur when (see equations (2.23)).

$$|\tfrac{1}{2}(\sigma_1 - \sigma_3)| \geqslant \pm S_{sc} \qquad (2.28a)$$

or using the above approximation at yield, that

$$|\sigma_1 - \sigma_3| \geqslant \pm S_p \text{ (or } S_c) \qquad (2.28b)$$

It is usual to apply this theory to ductile materials, good approximation being obtained between predicted and experimental values. The results are on the safe side independent of whether the principal stresses have like or unlike signs. However, it should be mentioned that if the principal stresses for a two-dimensional stress system are calculated using equation (2.19), and they have like signs, then the maximum shear stress must take account of the third principal stress being zero. This is inferred by the notation adopted.

Substituting for σ_1 and σ_2 from equation (2.19) into equation (2.23) gives the result

$$\left(\frac{\sigma_x - \sigma_y}{S_c}\right)^2 + 4\left(\frac{\tau_{xy}}{S_c}\right)^2 = 1 \qquad (2.29)$$

and in the special case when $\sigma_y = 0$, this reduces to

$$\left(\frac{\sigma_x}{S_c}\right)^2 + 4\left(\frac{\tau_{xy}}{S_c}\right)^2 = 1 \qquad (2.30)$$

2.3.4 Maximum Strain Energy Theory

The strain energy theory of elastic failure was first proposed by Beltrami, but often is attributed to Haigh. It is assumed that failure occurs when the strain energy per unit volume in the complex stress system is equal to some critical value, usually the strain energy per unit volume at the elastic limit in simple tension.

The strain energy per unit volume in simple tension is[1]

$$U_t = \frac{S_c \epsilon}{2} = \frac{S_c^2}{2E}$$

and the strain energy per unit volume for the complex stress case[1] is

$$U_c = \sum \frac{\sigma_1}{2E} [\sigma_1 - \nu(\sigma_2 + \sigma_3)]$$

$$= \frac{1}{2E} [\sigma_1^2 + \sigma_2^2 + \sigma_3^2 - 2\nu(\sigma_1 \sigma_2 + \sigma_1 \sigma_3 + \sigma_2 \sigma_3)]$$

Hence,

$$S_c^2 = \sigma_1^2 + \sigma_2^2 + \sigma_3^2 - 2\nu(\sigma_1\sigma_2 + \sigma_1\sigma_3 + \sigma_2\sigma_3) \qquad (2.31)$$

where in the special case, $S_c = S_p$.

Although results obtained using this theory give reasonably good correlation with experimental data for many ductile materials, it is not generally in common use.

2.3.5 Shear Strain Energy (or Distortion Energy) Theory

The shear strain energy theory, also known as the distortion energy theory, is due to von Mises and Hencky. According to this theory failure is assumed to occur when the shear strain energy for the complex stress system is equal to the shear strain energy at some critical value. Stated mathematically, failure is deemed to occur[1] when

$$(\sigma_1 - \sigma_2)^2 + (\sigma_2 - \sigma_3)^2 + (\sigma_3 - \sigma_1)^2 \geqslant 2S_c^2 \qquad (2.32)$$

Under static loading conditions the critical value S_c is usually taken as being equal to the proof or yield strength, S_p.

The distortion energy theory gives the best correlation between theoretical and experimental values for ductile materials, and even for brittle materials if the effect of stress concentration due to inclusions is considered. Of the failure theories presented, experimental and theoretical evidence suggests that the distortion energy theory is probably the most fundamental. In any event, only three failure theories find wide application, namely, the maximum principal stress (for brittle materials), and the maximum shear stress and distortion energy theories for ductile materials. In distinguishing between ductile and brittle materials, it is arbitrarily assumed that *ductile materials are those exhibiting a percentage elongation greater than 5 per cent*, although in reality each material must be assessed.

Substituting for σ_1 and σ_2 from equation (2.19) into equation (2.28) and transposing gives the result

$$\left(\frac{\sigma_x}{S_c}\right)^2 + \left(\frac{\sigma_y}{S_c}\right)^2 - \left(\frac{\sigma_x\sigma_y}{S_c^2}\right) + 3\left(\frac{\tau_{xy}}{S_c}\right)^2 = 1 \qquad (2.33)$$

which, for the special case, when $\sigma_y = 0$, reduces to

$$\left(\frac{\sigma_x}{S_c}\right)^2 + 3\left(\frac{\tau_{xy}}{S_c}\right)^2 = 1 \qquad (2.34)$$

Many problems in mechanical engineering design are concerned with components subjected to combined stresses, and the material presented in this section should enable an assessment of integrity to be made on a static basis. Figure 2.4 indicates a graphical representation of failure theories for use in static load design.

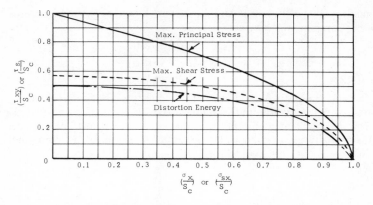

Figure 2.4 *Design curves representing failure theories*

2.4 Combined Fluctuating Stresses

The fatigue analysis of a component subject to combined fluctuating stresses requires first of all that the critical regions be located, and these are likely to be at points of high stress concentration, more particularly in a high stress field. Having located these critical regions, either by experience, methods of experimental stress analysis, or using theoretical predictions, the magnitude of the peak stresses and strains can be estimated. Once the peak stresses and strains have been calculated, it then remains to assess fatigue integrity under the action of the combined stresses.

In this section, the static theories of failure which have been discussed will be extended to the case of fluctuating stresses, although there is no reason to suppose that they should apply. However, the best justification for their use, from an engineering viewpoint, is that there is evidence that such an application enables the fatigue integrity of a component to be assessed with a fair degree of confidence.

For the case of a one-dimensional stress system subjected to a mean and alternating stress component, it was demonstrated in Section 1.8 that, introducing a reserve factor, RF, into the Goodman relationship, we could write

$$\text{RF} = \frac{S_u}{\sigma_m + c\sigma_{alt}} \qquad (1.17)$$

Further, if we compare this result with the usual definition for reserve factor under static loading, we may conclude that the denominator in equation (1.17) may be considered as an equivalent static stress having the same damaging effect as the actual defined fluctuating stress system. Consequently if the equivalent static stresses for a combined fluctuating stress system are calculated, and an appropriate failure theory incorporated in the analysis, this provides a means whereby the fatigue integrity of the component may be assessed.

Consider, then, a general co-planar stress system but, in this instance instead of the stresses being statically applied, let them be fluctuating periodically in phase. From the previous reasoning the equivalent static stresses in each direction, having the same damaging effect (or giving the same risk factor) as the mean and alternating components, may be written as

$$\sigma_{sx} = \sigma_{mx} + c_1 \sigma_{altx}$$
$$\sigma_{sy} = \sigma_{my} + c_2 \sigma_{alty} \tag{2.35}$$
$$\tau_s = \tau_m + c_3 \tau_{alt}$$

The mean and alternating components in the appropriate directions can be calculated from the maximum and minimum stress levels, using equations (1.8) and (1.9). The material and geometric constants c_1, c_2 and c_3 may be obtained from the ratio of the ultimate strength to the fatigue strength for the component in the direction concerned. In equation form,

$$c_1 = \frac{S_{ux}}{S'_{ax}}$$

$$c_2 = \frac{S_{uy}}{S'_{ay}} \tag{2.36}$$

$$c_3 = \frac{S_{su}}{S'_{sa}}$$

where S_{ux} and S_{uy} represent the ultimate tensile strength of the material in the x and y directions respectively, and S_{su} is the ultimate strength of the material in shear. In the absence of more specific information it may be reasonably assumed that $S_{su} = 0.75 S_u$. It is interesting to note at this stage that the effect of directional properties is included in equations (2.36) for both material variations, such as might be obtained with a forging, and also for type of loading and stress concentration. The modified fatigue strengths corresponding to the particular life of interest, and in the appropriate direction, are obtained by constructing a modified fatigue curve incorporating the factors discussed in Chapter 1. The principal stresses, referred to as the *equivalent principal stresses*, can now be calculated using the usual equation (2.19), thus,

$$\sigma_{1s} \text{ or } \sigma_{2s} = \tfrac{1}{2}(\sigma_{sx} + \sigma_{sy}) \pm \tfrac{1}{2}[(\sigma_{sx} - \sigma_{sy})^2 + 4\tau_s^2]^{1/2} \tag{2.37}$$

Substituting for σ_{sx}, σ_{sy} and τ_s from equation (2.37) we obtain

$$\sigma_{1s} \text{ or } \sigma_{2s} = \tfrac{1}{2}[(\sigma_{mx} + \sigma_{my}) + (c_1 \sigma_{altx} + c_2 \sigma_{alty})]$$
$$\pm \tfrac{1}{2}\{[(\sigma_{mx} - \sigma_{my}) + (c_1 \sigma_{altx} - c_2 \sigma_{alty})]^2 + 4[\tau_m + c_3 \tau_{alt}^2]\}^{1/2} \tag{2.38}$$

The results obtained using equations (2.38) and the further knowledge that σ_{3s} is zero, may now be incorporated in an appropriate failure theory to assess integrity. Consequently, if σ_x, σ_y and τ_{xy} are replaced by σ_{sx}, σ_{sy} and τ_s, i.e. the equivalent damage stresses, then equations (2.25), (2.29) and (2.33) may be used to assess integrity.

2.5 Methodical Procedure

The procedure for a multiaxial fatigue analysis in the HCF region may be summarised by the following steps:

(1) Estimate as accurately as possible the values for the factors K_f, K_s, C_s and C_L in each stress direction. If the geometry is not defined, assume that

$K_f = K_t$ and estimate an approximate value for K_t. At the feasibility stage, a useful value in the absence of other information is to assume that $K_t = 2$.

(2) Make a decision regarding the choice of material from which the component is to be produced (if this is not already known), and obtain an intrinsic fatigue curve for this material, either by predictive methods or using actual experimental data. Using the methods discussed in Chapter 1, produce modified fatigue curves for the components for each direction of stress, and corresponding to the desired life, determine the fatigue strengths S'_{ax}, S'_{ay}, S'_{sa}. The material and geometric constants (which also include a correction for anisotropy) can now be calculated using equations (2.36).

(3) Determine the nominal maximum and minimum stresses in the x, y and shear directions, and obtain the mean and alternating components using equations (1.8) and (1.9). The equivalent static stresses may now be determined by equations (2.35).

(4) Calculate the equivalent principal stresses using equations (2.37) or (2.38) and apply an appropriate failure theory to assess fatigue integrity or solve for the unknown quantity. If the material and geometry are completely defined, the unknown quantity will be the reserve factor. For a feasibility study, the unknown quantity will probably be the geometry, such as area, shaft diameter, second moment of area, etc., in which case it will be necessary to incorporate a value for K_t as stated in (1) above. Obviously, there are other variations on the same theme, such as having the geometry defined and using the procedure to select a material. This latter approach, however, is somewhat difficult in that c_1, c_2 and c_3 are initially unknown, and it therefore becomes necessary to make certain assumptions in order to obtain a solution. One such assumption which has had successful application in this sense is to assume initially that $c_1 = c_2 = c_3$. However, it is usually better to select a material based on other considerations, or possibly a somewhat crude static analysis incorporating a generous reserve factor (about 4 or 5), and then to modify the decision in accordance with the results of the fatigue analysis.

The decision regarding which failure theory to use is not always easy, since for those situations involving combined stresses with mean and alternating components, a tensile mean stress is more damaging than a compressive mean stress. It has been shown[6,7] that a modified maximum shear stress theory to account for the influence of a normal stress on the plane of failure and corrected for anisotropy, gives the best correlations with experimental data for smooth specimens. Since the distortion energy theory predicts equal damage independent of whether the mean stress is tensile or compressive, it is obviously not correct for all situations involving mean stresses. However, where the stresses are completely reversed, i.e. the mean stresses are zero, for ductile materials the distortion energy theory does give good correlation. For situations involving tensile mean stresses, either the distortion energy or the maximum shear stress theories may be used, and for situations involving compressive mean stresses, it is suggested that the mean stress be assumed zero and either the distortion energy or maximum shear theory be applied.

The most recent theory for fatigue failure under multiaxial stress–strain conditions is that due to Brown and Miller[8], which is based on the physical quantities that control fatigue crack growth, namely shear strain and tensile strain normal to the plane of maximum shear. It is claimed that contours of constant endurance may be represented graphically by the equation

$$\frac{\epsilon_1 - \epsilon_3}{2} = f\left[\frac{\epsilon_1 + \epsilon_3}{2}\right] \tag{2.39}$$

where ϵ_1 and ϵ_3 represent the maximum and minimum principal strains. The maximum shear strain, i.e. $(\epsilon_1 - \epsilon_3)/2$ is plotted as the abscissa and $(\epsilon_1 + \epsilon_3)/2$ is plotted as the ordinate. The theory illustrates the necessity to control bulk strain states, but this is obviously influenced by the geometry and type of loading imposed.

2.6 Combined Creep and Fatigue

The discussions in the present chapter have not so far considered the effect of environment, more particularly that of elevated temperature, although brief mention was made of this very important aspect in Chapter 1.

There are many situations requiring that a component should operate at elevated temperatures with and without fluctuating loads, such as aircraft components, land-based generating plant of all descriptions, nuclear pressure vessels and the like, and so on. Many of these situations require working to a code of practice, and allowable working stresses may be governed by legal requirements.

The first consideration to note is that the effect of temperature will, in general, be to alter the mechanical properties of materials from those normally obtained at room temperature, including the observation that metals which normally exhibit an endurance or fatigue limit at normal temperatures do not generally do so at high temperatures; secondly the effect of temperature may be to cause thermal stresses to be set up, due to either external restraint or to internal restraint caused by temperature gradients and differential expansion; and thirdly, if the temperature is sufficiently high (in excess of about 0.35–0.7 times the absolute melting point of the base material), progressive deformation of a material under constant load may occur, i.e. the material is subjected to *creep*. It is obviously necessary when designing components for operating in a particular set of environmental conditions to use the material properties corresponding to those conditions. The calculation of thermal stresses is not the subject of the present work and will therefore not be further discussed[5,9,10] but before proceeding to a consideration of a design method for combined fluctuating stresses at elevated temperature, it is necessary to consider briefly the phenomenon of creep.

Basic creep data is usually obtained by testing specimens under the condition of constant stress and temperature, either to produce definite amounts of deformation, or until actual rupture. These types of tests are known as *creep* and *creep rupture* tests respectively. Having obtained such basic data, it is then necessary to correlate the results of short term elevated temperature tests with long term performance at other temperatures, and certain predictive methods have been proposed, the most significant being the Larson–Miller[11] and the Manson–Haferd[12] methods.

The basic creep curve obtained under simple tension consists of the stages indicated in figure 2.5, the stage of usually greatest importance in design being the

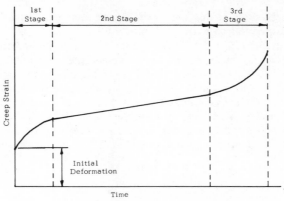

Figure 2.5 *Typical creep curve*

second stage. A mathematical model for secondary creep behaviour which has been
verified experimentally, is expressed in the following equation

$$C = \frac{\epsilon}{t} = B\sigma^n \tag{2.40}$$

where C = creep rate in tension, usually expressed as a % strain per 1000 h;
σ = applied tensile stress, MN/m^2 (lbf/in^2); and B and n are experimentally
determined constants whose values depend upon the material and operating
conditions. Typical creep data, plotted logarithmically, is shown in figure 2.6. The

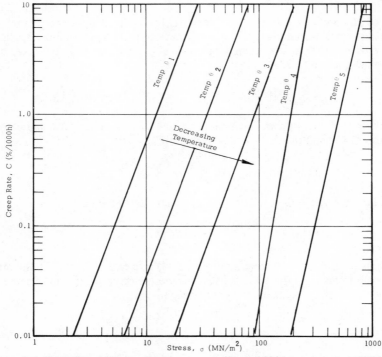

Figure 2.6 *Typical logarithmic plot of creep rate (% per 1000 h) against stress (MN m^{-2})*

exponent n is the slope of this line on the log-log plot, and B is the *creep rate* intercept corresponding to a stress of unity.

The data obtained from simple creep tests in tension can be used to predict creep rates under multiaxial loading conditions by the application of a failure theory. First of all it is necessary to write equations for the principal strains under plastic conditions, and this may be done by analogy with the equations for principal elastic strains, expressed by equations (2.22). It is usual to replace the reciprocal of the elastic modulus by a so-called *plasticity modulus* ϕ, and for consistency of volume, it is assumed that Poisson's ratio is equal to 0.5. Consequently, for creep conditions equation (2.22) may be written as

$$\epsilon_1 = \phi[\sigma_1 - 0.5(\sigma_2 + \sigma_3)]$$
$$\epsilon_2 = \phi[\sigma_2 - 0.5(\sigma_1 + \sigma_3)] \tag{2.41}$$
$$\epsilon_3 = \phi[\sigma_3 - 0.5(\sigma_1 + \sigma_2)]$$

Converting these equations to creep rate by dividing through by time, t,

$$C_1 = \frac{\epsilon_1}{t} = \frac{\phi}{t}\,[\sigma_1 - 0.5(\sigma_2 + \sigma_3)]$$

$$C_2 = \frac{\epsilon_2}{t} = \frac{\phi}{t}\,[\sigma_2 - 0.5(\sigma_1 + \sigma_3)] \tag{2.42}$$

$$C_3 = \frac{\epsilon_3}{t} = \frac{\phi}{t}\,[\sigma_3 - 0.5(\sigma_1 + \sigma_2)]$$

where C_1, C_2 and C_3 represent the creep rates in the directions of the three principal stresses. Assuming that the distortion energy theory of failure applies, i.e. that the three principal stresses may be replaced by an equivalent stress (say σ_e), then from equation (2.32)

$$\sigma_e = \frac{1}{\sqrt{2}}\,[(\sigma_1 - \sigma_2)^2 + (\sigma_2 - \sigma_3)^2 + (\sigma_3 - \sigma_1)^2]^{1/2} \tag{2.43}$$

Considering the case of simple tension, in which σ_2 and σ_3 are zero, and combining equation (2.43) and the first of equations (2.42)

$$\sigma_e = \sigma_1 = \frac{t}{\phi}\,C$$

and combining this with equation (2.40), noting that σ can be written as σ_e, gives

$$\frac{\phi}{t} = \frac{C}{\sigma_e} = B\sigma_e^{n-1} \tag{2.44}$$

Now substituting for (ϕ/t) from equation (2.44) into equations (2.42), the principal creep rates in terms of the principal stresses and the creep parameters B and n obtained from simple tension creep tests may be written

$$C_1 = B\sigma_e^{n-1}[\sigma_1 - 0.5(\sigma_2 + \sigma_3)]$$
$$C_2 = B\sigma_e^{n-1}[\sigma_2 - 0.5(\sigma_1 + \sigma_3)] \tag{2.45}$$
$$C_3 = B\sigma_e^{n-1}[\sigma_3 - 0.5(\sigma_1 + \sigma_2)]$$

Thus, if the principal stresses are calculated as discussed in Section 2.2, and the material parameters B and n are determined experimentally for the material of interest, the creep strain rates can be calculated using equation (2.45), and the values so obtained related to permissible values to assess the creep integrity.

For situations involving combined creep and fatigue, any analysis can only be, at the best, approximate. Nevertheless, such situations are encountered in design, and it is therefore necessary to have some method which enables an assessment to be made. The recommended procedure is to use a modified form of Goodman or master diagram, combining fatigue and creep data as indicated in figure 1.11. The limiting value for alternating stress corresponding to zero mean stress is the fatigue strength of the material at the temperature of interest; the limiting value of mean stress corresponding to zero alternating stress is either the creep rupture stress, or the stress corresponding to an acceptable creep elongation after a certain period of time. In equation form,

$$\frac{\sigma_{alt}}{S_a'} + \frac{\sigma_m}{S_c} = 1 \tag{2.46}$$

where S_a' is the modified fatigue strength of the component at the temperature of interest, and S_c is either the creep rupture stress or the stress corresponding to an acceptable creep elongation. This linear relationship is often found to be over-conservative, and an elliptical approximation has been suggested which, in equation form, may be written

$$\left(\frac{\sigma_{alt}}{S_a'}\right)^2 + \left(\frac{\sigma_m}{S_c}\right)^2 = 1 \tag{2.47}$$

The procedure for one-dimensional stress systems is straightforward, requiring only that the mean and alternating components be determined in the usual way from equations (1.8) and (1.9), but for combined stress systems, the approach is somewhat more complex. One suggestion which is intuitively reasonable, is to combine the methods of Section 2.4 with the equations for principal creep rates expressed by equations (2.45). This method would necessitate calculating the equivalent principal stresses using equations (2.37) or (2.38), and then substituting these values into equations (2.45) to obtain a solution for combined creep and fatigue integrity. This suggestion is not claimed as a rigorous one, and is open to criticism on the basis that it assumes constant equivalent principal stresses, whereas the actual system involves a different set of conditions. Even so, the method is one which should prove useful in many circumstances, and it is one which is likely to appeal to the designer.

A second approach is to determine a so-called equivalent mean stress and an equivalent alternating stress, and use one of the equations (2.46) or (2.47) to obtain a solution. The equivalent mean and alternating stresses are obtained using an appropriate theory of failure. Thus, for a system in which the three principal stresses vary periodically in phase between maximum and minimum values, the equivalent mean and alternating stress components would be, using the distortion energy theory,

$$\sigma_{me} = \frac{1}{\sqrt{2}} [\sigma_{m_1} - \sigma_{m_2})^2 + (\sigma_{m_2} - \sigma_{m_3})^2 + (\sigma_{m_3} - \sigma_{m_1})^2]^{1/2} \tag{2.48}$$

$$\sigma_{alte} = \frac{1}{\sqrt{2}} [(\sigma_{alt_1} - \sigma_{alt_2})^2 + (\sigma_{alt_2} - \sigma_{alt_3})^2 + (\sigma_{alt_3} - \sigma_{alt_1})^2]^{1/2} \tag{2.49}$$

where σ_{me} and σ_{alte} are defined as the equivalent mean and alternating stresses respectively, and σ_{m1}, σ_{alt1}, σ_{m2}, σ_{alt2}, and σ_{m3}, σ_{alt3} represent the mean and alternating components respectively in the directions of the principal stresses. Consequently, these mean and alternating components may be represented on a modified Goodman diagram or master diagram, or equations (2.46) and (2.47) may be used to give a solution. However, this requires that a modified fatigue curve be assumed in order to estimate a value for S'_a, and the question is immediately posed as to how such a curve may be derived. In the absence of more specific data and information relating to particular applications, it is suggested that the first approach to combined creep and fatigue under fluctuating stresses be used in design situations.

The further problem of combined creep and fatigue of nonconstant amplitude must be dealt with by an appropriate damage theory, and despite the criticisms, a cumulative damage law will usually be applied. Expressed mathematically, the concept of life-fraction-damage accumulation may be written as[13]

$$\sum \left(\frac{n}{N}\right)_j + \sum \left(\frac{t}{T}\right)_k = D \tag{2.50}$$

where D = total damage factor; n = applied cycles of load condition j; N = cycles to failure at load condition j; t = time duration of load condition k; T = failure time at load condition k. Since the first term in equation (2.50) may include some creep damage at low cyclic rates, it is suggested[13] that this formulation does not really effect the separation of plasticity damage from creep damage, and the strain range partitioning proposed by Manson and his co-workers[14] might eventually become more useful.

2.7 Concluding Remarks

The discussions presented in this chapter are aimed at providing reasonable working design procedures for assessing fatigue integrity of components subjected to combined fluctuating stresses. No attempt has been made to consider the fundamental aspects of the actual mechanism of failure, the emphasis being rather on the application of basic fatigue data to design. It is recognised that some of the methods may not be considered very sophisticated, but until the question of how they may be improved can be answered with confidence it is hoped that they will prove useful in many design situations.

Other methods of dealing with combined fluctuating stresses have been suggested[15-18] for the case where the stresses fluctuate in phase, and also for the unusual circumstance where combined out of phase fluctuating stresses are encountered[19]. Nevertheless, the equivalent damage concept used in conjunction with an appropriate failure theory as discussed, should prove adequate for most design applications.

The treatment of creep and creep—fatigue interactions is of necessity somewhat superficial, and only sufficient coverage has been included to enable the engineer to approach the problem of combined fluctuating stresses at elevated temperatures. A useful review of creep—fatigue interactions has been produced by Ellison[20] and a discussion on service life predictions can be found in Ellison and Smith[21].

References

1. Duggan, T. V., *Stress Analysis and Vibrations of Elastic Bodies*, Iliffe, London (1963)
2. Duggan, T. V., *Applied Engineering Design and Analysis*, Iliffe, London (1970)
3. Korn, G. A. and Korn, T. M., *Mathematical Handbook for Scientists & Engineers*, McGraw-Hill, New York (1961)
4. Pipes, L. A., *Applied Mathematics for Engineers and Physicists*, 2nd edn, McGraw-Hill, New York (1958)
5. Faupel, J. H., *Engineering Design*, Wiley, New York (1964)
6. Findley, W. N., Coleman, J. J. and Hanley, B. C., Theory for Combined Bending and Torsion Fatigue with Data for SAE 4340 Steel, *Proc. Intern. Conf. on Fatigue of Metals*, p. 150, Inst. of Mech. Engrs, London (1956)
7. Findley, W. N., A Theory for the Effect of Mean Stress on Fatigue of Metals Under Combined Torsion and Axial Load or Bending, *J. Engineering for Industry*, Trans. of ASME, November (1959)
8. Brown, M. W. and Miller, K. J., A Theory for Fatigue Failure Under Multiaxial Stress-Strain Conditions, *Proc. Inst. Mech. Engrs,* **187**, 65/73, 745–755 (1973)
   (See also Discussion 1 this paper, pp. D 229–D 244)
9. Boley, B. A. and Weiner, J. H., *Theory of Thermal Stresses*, Wiley, New York (1960)
10. Gatewood, B. E., *Thermal Stresses*, McGraw-Hill, New York (1957)
11. Larson, F. R. and Miller, J., Time Temperature Relationships for Rupture and Creep Stresses, *Trans. ASME*, **74**, 756 (1952)
12. Manson, S. S. and Haferd, A. M., A Linear Time Temperature Relation for Extrapolation of Creep and Stress Rupture Data, *NACA Tech. Note 2890* (1953)
13. Langer, B. F., Design Aspects of Elevated Temperature Technology, *Int. Conf. on Creep and Fatigue in Elevated Temperature Applications*, Inst. of Mech. Engrs (1974)
14. Manson, S. S., Halford, G. R. and Hirschberg, M. H., Creep-fatigue Analysis by Strain-range Partitioning, *NASA Tech. Memor., NASA-TM-X-67838* (1971)
15. Juvinall, R. C., *Engineering Considerations of Stress, Strain and Strength*, McGraw-Hill, New York (1967)
16. Marin, J., Interpretation of Fatigue Strengths for Combined Stresses, *Proc. Intern. Conf. on Fatigue of Metals*, 184, Inst. Mech. Engrs, London (1966)
17. Stulen, F. B. and Cummings, H. N., A Failure Criterion for Multiaxial Fatigue Stresses, *Proc. Am. Soc. for Testing and Materials*, **54**, 822 (1954)
18. Ellison, E. G. and Andrews, J. M. H., Biaxial Cyclic High Strain Fatigue of Aluminium Alloy RR58, *J. Strain Anal.*, **8** (3), 209 (1973)
19. Little, R. E., Fatigue Stresses for Complex Loadings, *Machine Design*, **38**, 145 (1966)
20. Ellison, E. G., A Review of the Interaction of Creep and Fatigue, *J. Mech. Engng Sci.*, **11** (3), 575–611 (1969)
21. Ellison, E. G. and Smith, E. M., Predicting Service Life in a Fatigue Creep Environment, *ASTM STP 520* (1973)

Cyclic Material Behaviour

3.1 Introduction

It is now generally accepted that the fatigue process is one of energy conversion, and the energy responsible for the formation of an engineering crack is that of plastic flow. As a consequence, it might be thought that limiting the amount of plastic flow will have the effect of increasing the fatigue life of a structure. In general, this conclusion is correct, provided that it is limited to describing only the crack formation stage.

Engineering metals and alloys are made up of an aggregate of crystals, each of which usually has the atoms arranged in a definite pattern. Since a crystal has a directional structure it possesses directional properties, but these directional properties are inhibited in a real material where the random orientation of millions of crystals leads to a cancellation of the directional properties on a macroscopic scale; this situation gives rise to the usual assumptions of a material being homogeneous and isotropic. Thus, it is seen that a material has both microscopic and macroscopic properties, the nominal stress—strain response being usually taken as representing the mechanical property of the material on a macroscopic scale. Since every macroscopic element is made up of infinitesimal elements having a range of values of stress—strain response, a real continuous medium will exhibit both inhomogeneity and anisotropy. Since fatigue damage is usually initiated at the microscopic level the heterogeneous nature is of importance, and variations in microstructural properties due to differences in grain size, inclusions, anisotropy, orientation, microresidual stress, and so on, need to be considered[1].

3.2 Formation of Fatigue Cracks

Crack initiation will occur in a real component or structure at some critical region corresponding to high stress concentration, such as a discontinuity due to change in section; in the heat affected zone of a welded joint; or due to poor surface finish or some other topographical feature. In the case of a smooth test specimen, nucleation will occur at some local weakness or structural defect.

Multinucleation of fatigue cracks is usually found in test specimens, with eventually one nucleation source developing into a dominant fatigue crack which subsequently may grow to critical conditions. In components, however, there is some evidence that single nucleation sources predominate[2], and the consequential scatter associated with fatigue life may be reduced.

46

The prediction of crack formation is of immense importance, since if this can be achieved with a high confidence level the risk of failure is obviously minimised, and the subsequent life spent in fatigue crack propagation may be assessed using fracture mechanics concepts.

For a particular material there is an inverse relationship between the absolute stress or strain amplitude and the proportion of cycles spent in crack formation. If a material is subjected to very high cyclic stresses or strains, the proportion of cyclic life spent in forming a crack, as distinct from propagating the crack to critical conditions, will be very small, typically in the region of 10 per cent or less, and the actual number of cycles involved will also be small, say less than 50 000. On the other hand, for relatively low stresses or strains, the proportion of cyclic life involved in forming a crack may be 90 per cent or more of the total, and the number of cycles involved may be in excess of one million. If the visible appearance of a fatigue crack cannot be tolerated, then predicting fatigue crack formation may be equally important for both high or low stress or strain cycling.

It is evident from the above discussion that absolute stress or strain amplitude influences the formation of a fatigue crack, and it might be expected that this would be related to the accumulation of plastic strain energy.

Quite obviously the accumulation of plastic strain energy with cycles is larger for higher strain amplitudes than for low strain amplitudes, and in the HCF region the microstructural properties of the material are likely to be much more significant. In fact, it has been proposed by Esin[1] that in the HCF region fatigue damage is related to the accumulation of microplastic strain energy, and since microplastic flow is a random and microstructure sensitive property, a statistical approach is used. This is consistent with the stochastic nature of fatigue. The chief characteristic of HCF is that fatigue damage tends to spread progressively across the whole grain, and at a macroscopic level no significant cyclic strain hardening or softening is observed.

In the intermediate region, the strain amplitudes are such that again only microplasticity is involved, and the usual linear relationship between stress and strain applies, enabling strain amplitude to be replaced by stress amplitude, thus producing the usual $S-N$ curve. The amount of scatter in test data is reduced considerably with increasing strain amplitude, indicating less dependence on microstructural properties and possibly more on bulk properties.

If the strain amplitude is increased to the level where macroscopic plasticity is produced, i.e. a closed hysteresis loop is developed, there is no longer a linear relationship between stress and strain, and significant cyclic strain hardening or softening may occur; this is the true LCF region.

3.3 Low Cycle Failure

Low cycle fatigue was defined in Chapter 1 as those combinations of strain and number of cycles during which considerable macroplasticity occurs. If this definition is accepted, it necessitates that plastic strain energy must accumulate with cycles until a certain critical value is reached, at which stage a crack may form. Subsequent cycling may cause propagation of the crack until it reaches critical conditions, and fast fracture will then occur.

If low cycle failures are to be minimised, it is necessary that the fundamental mechanisms which cause failure are understood. Only then can components be designed to withstand low cycle failure with a high degree of confidence. In terms of cyclic life, the division between low cycle and high cycle failure is usually considered to be somewhere between about 10 000 and 50 000 cycles, although this division is an arbitrary one and is dependent upon the relative magnitudes of plastic and elastic strains.

Failures in the low cycle region are possible in many dynamic situations where the stresses accompanying cyclic loading exceed the macroscopic yield strength. Typical situations where this condition is possible are found in nuclear pressure vessels, modern steam turbines and aero-engine components where local discontinuities and geometrical concentration features cause local stresses to be set up which may (on an assumed elastic analysis) be considerably in excess of the macroscopic yield strength. Under these conditions, stress is not a particularly meaningful quantity, and it is more useful to think in terms of strain. In most practical designs the local plastic strains will usually be sufficiently contained to limit the plastic zone to only a small region, i.e. the ratio of plastically to elastically strained material will usually be quite small. Under these circumstances, even though the component as a whole may be subjected to constant load cycling (due, for example, to start-ups and shut-downs, which may or may not involve temperature fluctuations), the material in the vicinity of a concentration feature will have a cyclic stress—strain response quite different from that of the bulk material. This local behaviour will be dependent upon a number of factors, such as the relative magnitude of plastically to elastically strained material, the strain distribution, the material's cyclic strain hardening or strain softening characteristics, and the effect of environment, more particularly that of temperature. Further, even though the component may be subjected to constant amplitude cyclic loading, the material which is locally plastic will experience a variation in strain range with cycles[3]. It is the behaviour of this local material, more particularly the way in which plastic strain or plastic strain energy accumulates with cycles, which governs the mechanism responsible for low cycle failure.

As already mentioned, low cyclic failure requires a cyclic accumulation of plastic strain or plastic strain energy, and this can occur in a number of different ways. It may be brought about by mechanical load cycling; by thermal cycling or by a combination of both mechanical and thermal cycling. The mechanism may be further complicated by unstable material behaviour (so-called 'second order plasticity'), the presence of 'follow-up loads', and the incorporation of stress relaxation due to hold periods at elevated temperatures.

The manner in which cyclic plastic strain may be introduced into a component is sometimes referred to as either strain limiting or load limiting deformation depending on whether deformation of the material causes the deformation loads to be relieved (i.e. the strains are self limiting), or whether the restraining forces in the material increase to balance the applied load.

Failure mechanisms in the low cycle region may be, broadly speaking, divided into three categories, although further subdivision on a more detailed level is possible. These three categories may be described as:

(1) low cycle fatigue (LCF) failures;
(2) deformation type failures; and
(3) combined LCF and deformation type failures.

3.4 High Strain Low Cycle Fatigue

In many components subjected to mechanical load or thermal cycling it is likely that, at a typical concentration feature, local plasticity due to yielding will be obtained, at least for the first half cycle. If continued cycling does not produce any plastic strain, i.e. a so-called 'elastic shakedown' condition is obtained, the material will behave elastically and low cycle failure would not be expected to occur. This is illustrated in figure 3.1 where OA represents the first half cycle, and AB the second

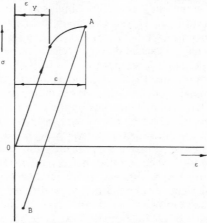

Figure 3.1 *Local stress—strain behaviour for first cycle* $(\epsilon < 2\epsilon_y)$

half, for the material in the vicinity of a concentration feature. In this illustration, yielding does not occur during the second half cycle, and during subsequent cycling, the material will simply be strained elastically up and down the line **AB**. On the other hand, if the plastic strain obtained during the first half cycle is sufficiently great (i.e. in excess of twice the yield strain), some yielding would be expected on the second half cycle, as shown in figure 3.2. Whether subsequent

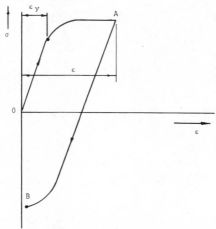

Figure 3.2 *Local stress—strain behaviour for first cycle* $(\epsilon > 2\epsilon_y)$

cyclic loading continues to produce cyclic plasticity will depend upon the magnitude of the cyclic load and the material characteristics. In such instances, although the component as a whole may be subjected to constant load cycling (assumed), the material in the vicinity of a concentration feature will experience a condition approaching that of constant strain cycling. This infers that the strain range $\Delta\epsilon$ remains approximately constant, but the magnitude of the residual strain will depend upon the relative volume of plastically to elastically strained material. With continued cycling, a stable hysteresis loop may be developed, and an essentially constant strain range condition obtains as indicated in figure 3.3; the magnitude of the mean strain ϵ_m depends upon the degree of plastic penetration.

In order to obtain true LCF failure, it is necessary for yielding to occur on both the first and second half of each and every cycle, this phenomenon being very much dependent upon the elastic restoring forces for the bulk material acting on the local plastic region. It is easy to see that if a closed hysteresis loop is obtained cyclic plasticity will accumulate. Further, when the accumulation of plastic strain energy reaches a certain critical value, depending on the material, a crack will develop. The true LCF mechanism, indicated by the closed hysteresis loop shown in figure 3.3, does not take account of the effect of complicating factors which might influence the shape of the hysteresis loop and consequently the mechanism of failure. It is perhaps worthy of note to mention that, based on the static stress—strain curve, a low cycle failure is sometimes obtained under conditions which appear to correspond to elastic shakedown conditions, i.e. although yielding occurs on the first half cycle, the static properties indicate that subsequent cycling would not be expected to cause further cyclic plasticity. The fact that low cycle failure occurs at all, is indicative of cyclic plasticity, and further study may indicate that the

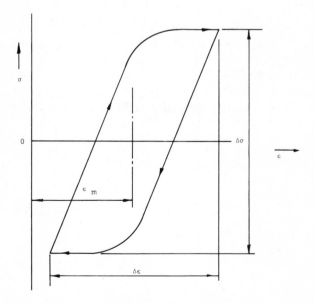

Figure 3.3 *Stable hysteresis loop for constant strain hypothesis*

material of interest is one which exhibits cyclic strain softening. This demonstrates the possible danger in using the static stress—strain curve for predicting cyclic behaviour, and emphasises the necessity of relating cyclic behaviour in components to the cyclic characteristics of the material.

The constant strain hypothesis is approached when components are subjected to thermal cycling, although the idealised model will be complicated by cyclic strain hardening characteristics of the material, and the fact that due to the continuously changing temperature the stress—strain response will also be changing continuously[4]. Nevertheless, neglecting complicating factors, the LCF failure mechanism may be expected to apply to components subjected to thermal cycling (provided that sufficient plastic strains are involved), the thermal strain being converted to mechanical strain; using a constant strain fatigue curve for the correct temperature, the cyclic life may be approximated.

3.5 Presentation of LCF Data

It is necessary to establish an acceptable definition for failure. In terms of crack formation life, this may be defined as the development of an engineering crack, which, for practical purposes is one which can be detected using low power magnification (say x25). A typical surface crack defined in this way will be in the region of about 0.5 mm long and 0.15 mm deep.

The formation of an engineering crack will involve nucleation, Stage I crack propagation and also some Stage II crack propagation. This statement is true both for components and also for laboratory test specimens, and a worthy aim is that of attempting to predict the behaviour of components from tests on simple test pieces. Since the failure of a simple test piece can be defined to correspond to the development of an engineering crack, this information may then be used to predict failure in a component produced from the same material and subjected to an identical environment.

A fundamental approach to assessing the number of cycles required to produce an engineering crack would necessitate the ability to determine precisely the division between nucleation and Stage I and Stage II crack propagation. Whilst for a limited number of situations it might be possible from a diagnostic viewpoint to use such an approach, the prognostic situation is much more difficult. The use of electron microscopy to count striations on a fracture surface is relatively commonplace and can provide useful information for failure diagnosis, but this is of only limited value to the designer. Consequently, despite the obvious criticisms, attempts have been made to develop methods which enable crack formation in components to be predicted from data obtained from simple test specimens, without any attempt to distinguish between the relative proportions of cyclic life spent in nucleation and Stage I and Stage II crack propagation. However, the proportion of Stage II crack propagation is likely to be very small in producing an engineering crack as previously defined. In a later section the prediction of Stage II crack propagation will be discussed using fracture mechanics concepts. Thus, knowing the number of cycles to produce an engineering crack and the subsequent number of cycles to grow that crack to some 'critical' condition, the fatigue integrity of a component may be assessed.

Fatigue tests may be conducted under conditions of either constant load (or stress) cycling or constant strain or deformation cycling. In the LCF region, constant strain cycling is considered to be more meaningful and therefore more useful when assessing the integrity of components. It would be misleading, however, to infer that all LCF testing is conducted using strain control and there are many workers who prefer load control because of its ready association with stress.

It is appropriate to consider briefly, the way in which an intrinsic constant strain control LCF curve is obtained. Two modes of strain control may be adopted, i.e. either longitudinal or diametral, each of which have certain advantages and disadvantages. If longitudinal strain control is used, this necessitates a specimen having a parallel gauge length over which a longitudinal extensometer is fitted, whereas for diametral strain control an hourglass specimen is used with a diametral extensometer fitted at the minimum section diameter.

In a study to investigate the significance of various parameters on high strain fatigue, Rigg[5] used hourglass specimens under diametral strain control, and discusses thoroughly the relative merits of the two modes. At elevated temperature, Rigg argues that an hourglass specimen makes it easier to control temperature to a high degree of accuracy where the specimen is to remain visible throughout the testing. With a parallel gauge length specimen a uniform temperature distribution is essential along the gauge length if localisation effects are to be avoided. An hourglass specimen has greater resistance to buckling than has a parallel gauge length specimen, a factor which may be important at high strains and low cyclic lives. A further advantage associated with hourglass specimens is that the failure origin is confined to the minimum section (or close to it), whereas with a parallel gauge length the nucleation source may occur anywhere along the parallel section.

Miller[6], on the other hand, has suggested several criticisms of the hourglass specimen. In the first instance he suggests that deformation at the minimum section may ultimately be so inhomogeneous as to necessitate applying correction factors to the true stress—true strain curve. Secondly, since the straining rate varies throughout the plastically deformed zone, varying degrees of interaction between time and cycle dependent processes may be permitted. Thirdly, it is suggested that geometry influences the results due to strain localisation effects. Rigg has attempted to answer these criticisms for his particular programme, but clearly each case must be judged on its merits. In any event, the usual method of presenting LCF test data is to use longitudinal strain range as the ordinate, and if diametral strain is used in the testing, this must be converted to longitudinal strain for design purposes. Rigg[5] derives a relationship between longitudinal and diametral strain, so that longitudinal strain range $\Delta \epsilon_L$ may be expressed in terms of the diametral strain range $\Delta \epsilon_D$ and stress range $\Delta \sigma$.

Expressed mathematically,

$$\Delta \epsilon_L = (\Delta \epsilon_D / \nu_p) + (\Delta \sigma / E)[1 - (\nu_e / \nu_p)] \tag{3.1}$$

where ν_e = Poisson's ratio in the elastic region; ν_p = Poisson's ratio in the plastic region; E = elastic modulus.

Assuming $\nu_p = 0.5$ in the plastic region, then equation (3.1) becomes

$$\Delta \epsilon_L = 2 \Delta \epsilon_D + (\Delta \sigma / E)(1 - 2 \nu_e) \tag{3.2}$$

In equations (3.1) and (3.2) $\Delta\sigma$ and $\Delta\epsilon_D$ are the measured variables, and E is found for the temperature and strain rate of the fatigue test. From a stable hysteresis loop of $\Delta\sigma$ against $\Delta\epsilon_D$, the slope of the elastic portion is E/ν_e, so that if E is known then ν_e can be calculated and all terms on the right hand side of the equation (3.2) are known, enabling $\Delta\epsilon_L$ to be determined. It may be observed from equation (3.2) that for a fixed value of diametral strain range, the longitudinal strain range will increase or decrease with cyclic strain hardening or softening, i.e. with variations in $\Delta\sigma$. This variation in $\Delta\sigma$ with cycles will now be discussed by describing the manner in which a cyclic stress—strain curve may be obtained.

The cyclic stress—strain curve represents the cyclic, as opposed to the monotonic, properties of a material. Such cyclic material behaviour representation may be obtained using several methods, but the most conventional method is that proposed by Manson[7]. In this method a series of tests is carried out under conditions of fully reversed constant strain, and the corresponding stress amplitude is monitored. Corresponding to a particular value of longitudinal strain amplitude, the stress amplitude will vary with cycles as shown in figure 3.4 for a cyclically

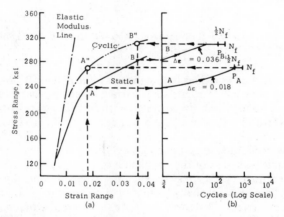

Figure 3.4 *Static and cyclic stress characteristics for cyclic strain hardening material: (a) static and cyclic stress—strain characteristics; (b) stress range as a function of applied cycles (Manson)*

strain hardening material, and figure 3.5 for a cyclically strain softening material. It is observed that, depending upon the strain amplitude and material, the variation of stress amplitude shakes down to a saturation level corresponding to a stable hysteresis loop; for very high strains and short cyclic lives, however, a saturation level may not be reached.

For the majority of materials, it seems that the saturation stress is approached after a number of cycles corresponding to the half life condition. Consequently, a point on the cyclic stress—strain curve is obtained by correlating the saturation stress amplitude, or the stress amplitude at the half life if saturation does not occur, with the appropriate value of longitudinal strain amplitude. For a material exhibiting cyclic strain softening this could theoretically give rise to a discontinuity in the cyclic curve at the monotonic yield strain. Since, however, the macroscopic yield strain may be represented by the mean of a distribution of micro-yield strains,

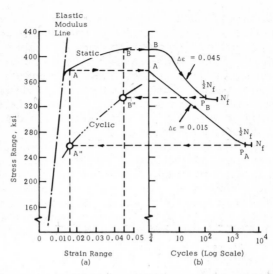

Figure 3.5 *Static and cyclic stress characteristics for cyclic strain softening material: (a) static and cyclic stress—strain characteristics, (b) stress range as a function of applied cycles (Manson)*

some elements will yield at strains below the microyield, and the discontinuity is consequently not observed. A cyclic stress—strain curve may be obtained in the manner described from conventional strain controlled fatigue tests to obtain an LCF curve, provided that the stress amplitude corresponding to each strain level is monitored.

An alternative technique for producing a cyclic stress—strain curve, using only one or two specimens, is that of the so-called incremental step test[8]. This method enables an approximate cyclic stress—strain curve to be obtained by subjecting a test specimen to blocks of gradually increasing then decreasing strain amplitude while under strain control. Hysteresis loops are recorded continuously during the course of each block, and the locus of the tips of the superimposed hysteresis loops traces the cyclic stress—strain curve. Figure 3.6 (from Rigg[5]) illustrates two sets of hysteresis loops so obtained for a martensitic creep resisting stainless steel (FV 535) at 500°C, and figure 3.7 shows the monotonic and cyclic stress—strain curves for the same material and temperature.

The basic data required by the designer is the intrinsic LCF curve and the cyclic stress—strain curve, and in the predictive methods to be discussed presently, it will be assumed that such data is either available or can be estimated with reasonable confidence. Considerable discussion could be included concerning the cyclic behaviour characteristics of specific materials, but since this is not the purpose of the present discussion the references available[7−10] should be consulted. It is sufficient to say that, under cyclic loading, materials attempt to revert to a stable condition by either cyclically strain hardening or softening. This characteristic appears to be associated with the prior mechanical work and heat treatment conducted on the material. The degree of cyclic strain hardening or strain softening varies considerably, some materials being almost neutral in this respect. Environ-

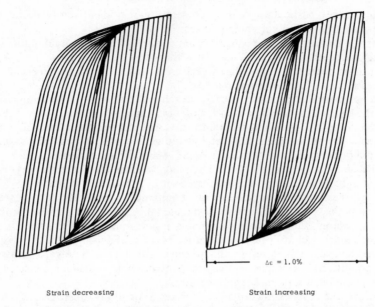

Strain decreasing   Strain increasing

Figure 3.6 *Hysteresis loops recorded during incremental step tests on FV 535 at 500°C*

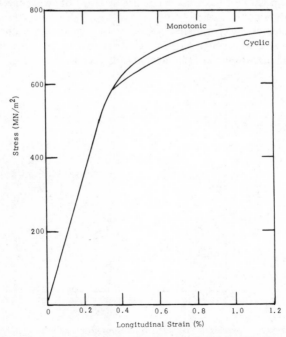

Figure 3.7 *Monotonic and cyclic stress—strain curve for FV 535 at 500°C*

mental influences, particularly that of temperature, may significantly affect the
cyclic strain hardening or softening characteristics.

Under stress control, as opposed to strain control, the cyclic strain hardening or
softening characteristics are evidenced by a change in the strain range, i.e. for a
constant stress range the strain range will decrease for a cyclic hardening material
and increase for a cyclic softening material.

As an approximate guide to the cyclic dependent hardening or softening of
metals, the initial or monotonic strain hardening exponent, n, may be used. For n
within the range 0.10–0.20, in general the material will exhibit only small cyclic
adjustments. Above this range the metal would be expected to cyclically strain
harden and below to cyclically strain soften[8]. Manson[4] has suggested a more
sophisticated approach using a cyclic hardening coefficient to determine cyclic
hardening or softening characteristics, based on LCF predicted behaviour.

In order to estimate the fatigue life of a component it is necessary to relate the
conditions at the critical region to known material behaviour, and allow for the
influencing factors. In brief, then, a desirable approach would be to calculate the
strain range in the vicinity of a concentration feature, and using an LCF curve, to
determine the cyclic life corresponding to the calculated strain range. This assumes
that the conditions in a component can be represented by tests on plain specimens
of the type used in obtaining the fatigue data.

It is usual when expressing fatigue data in terms of strain range versus cycles, to
use total strain range as the ordinate, as indicated in figure 3.8. However, since
every value of total strain is made up of an elastic component and a plastic
component, it is possible to express cyclic life in terms of either the elastic or the
plastic strain component.

There is considerable experimental support[7] to suggest that both the elastic and
plastic strain range components, when plotted against cycles, give approximately
straight lines on logarithmic co-ordinates, as indicated in figure 3.8. Thus, expressed

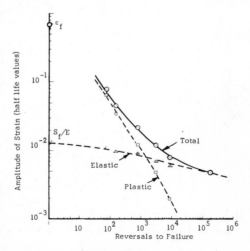

Figure 3.8 *Fatigue properties of 7075-T6 aluminium alloy (points are average of several specimens) (Sander[8])*

mathematically, the total strain range can be related to the fatigue life, N_f, by an equation of the form

$$\Delta\epsilon_T = C_p N_f^{\alpha_1} + C_e N_f^{\alpha_2} \tag{3.3}$$

where a_1 and α_2 represent the slopes of the plastic and elastic lines respectively on logarithmic co-ordinates, and C_p and C_e represent the strain range corresponding to the plastic and elastic intercept for one cycle. Equation (3.3) cannot be solved explicitly for N_f, and it is therefore convenient to express cyclic life in terms of the elastic or plastic strain component only. Thus, the plastic line may be represented by the equation

$$\Delta\epsilon_p = \epsilon_f' N_f^{\alpha_1} \tag{3.4}$$

or

$$N_f = \left(\frac{\Delta\epsilon_p}{\epsilon_f'}\right)^{1/\alpha_1} \tag{3.4a}$$

where ϵ_f' is defined as the fatigue ductility coefficient and α_1 as the fatigue ductility exponent.

It is observed that the plastic and elastic lines intercept (figure 3.8), and the cyclic life corresponding to this point is defined as the transition life. Below this transition life the plastic strain range dominates and it is convenient, therefore, to use plastic strain range as the describing parameter, i.e. equations (3.4) and (3.4a) are used to describe fatigue behaviour in this region. For cyclic lives in excess of the transition life, i.e. in the intermediate and HCF region, it is the elastic strain range which dominates, and so this is usually used as the describing parameter. The elastic line is usually expressed in the form

$$\Delta\epsilon_e = \left(\frac{S_f'}{E}\right) N_f^{\alpha_2} \tag{3.5}$$

or

$$N_f = \left(\frac{\Delta\epsilon_e E}{S_f'}\right)^{1/\alpha_2} \tag{3.5a}$$

where S_f' is defined as the fatigue strength coefficient, and α_2 is often referred to as Basquin's exponent[9]. As already mentioned, however, in the intermediate and HCF regions fatigue data is usually expressed in terms of conventional stress versus life fatigue curves and the methods discussed in Chapters 1 and 2 are applicable.

Ideally, material data should be obtained experimentally using constant strain control, but this is a costly and time consuming exercise. From a design point of view, particularly at the feasibility stage, an approximation is often of considerable value in assessing the design before time and effort is invested in a test programme. For this reason, it is desirable to consider the possibility of predicting an LCF fatigue curve using readily available material data. Two such methods which have been proposed, based upon empirical data, are the *method of universal slopes* and the so-called *four-point correlation method*. The universal slopes approach utilises the observation that, in low cycle fatigue a logarithmic plot of plastic strain against cycles for most metals results in a more or less constant slope value (usually between about −0.5 and −0.7).

The four-point correlation method uses the straight line relationships obtained for elastic and plastic components, and defines two points on each, as indicated in figure 3.9. P_1 and P_2 represent points on the elastic line and P_3 and P_4 points on the plastic line. Based upon the analysis of a large amount of experimental data[7], Manson has suggested the following empirical values. On the elastic line:

P_1, elastic strain range at $\frac{1}{4}c$, $\Delta\epsilon_e = 2.5\left(\dfrac{S_f}{E}\right)$ (3.6)

P_2, elastic strain range at 10^5c, $\Delta\epsilon_e = 0.9\left(\dfrac{S_u}{E}\right)$ (3.7)

On the plastic line:

P_3, plastic strain range at $10c$, $\Delta\epsilon_p = \frac{1}{4}D^{3/4}$ (3.8)

P_4, plastic strain range at 10^4c, $\Delta\epsilon_p = \dfrac{0.0132 - \Delta\epsilon_e^*}{1.91}$ (3.9)

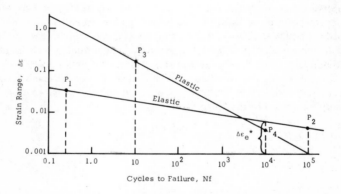

Figure 3.9 *Manson four point correlation prediction*

In equation (3.6), the fracture strength, S_f, is defined as the load at fracture divided by the final cross-sectional area. In the absence of precise information, the following approximation is suggested

$$S_f = (1 + D)S_u$$ (3.10)

where

$$D = \text{logarithmic ductility} = \ln 1/(1 - \text{RA})$$ (3.11)

and S_u = ultimate strength; RA = reduction in area.

In calculating the plastic strain range at 10^4c, it is first necessary to draw the elastic line and obtain the elastic strain range at 10^4c, i.e. $\Delta\epsilon_e^*$; this value is then used in equation (3.9). Thus, two straight lines may be drawn on logarithmic co-ordinates; the elastic line through points P_1 and P_2 and the plastic line through points P_3 and P_4. Hence, the sum of these individual components will give the total strain range corresponding to any particular cyclic life, and the curve for total strain range versus cycles can be obtained.

The Manson—Coffin relationship, given by equation (3.4), does not provide for the possible existence of an endurance limit, and Manson[4] has suggested modifying the equation to the more general form

$$(\Delta\epsilon_T - \Delta\epsilon_a) = \epsilon_f' N_f^{\alpha 1} \qquad (3.12)$$

where $\Delta\epsilon_T$ = total strain range = $\Delta\epsilon_p + \Delta\epsilon_e$; $\Delta\epsilon_a$ = endurance strain range.

A further modification, to allow for the effect of mean strain, has been suggested by Sachs et al.[11], which subtracts the mean strain ϵ_m from the fatigue ductility coefficient. Introducing this modification into equation (3.12) and transposing gives

$$N_f = \left(\frac{\Delta\epsilon_T - \Delta\epsilon_a}{\epsilon_f^1 - \epsilon_m}\right)^{1/\alpha 1} \qquad (3.13)$$

If the mean strain is small compared with the value of the fatigue ductility coefficient, ϵ_f', its influence will be insignificant.

Now it will be observed that equation (3.13) enables an estimate of cyclic life to be obtained if the strain conditions in a component can be calculated. However, since equation (3.13) is derived from fatigue data on test specimens, this statement infers that a component having a similar strain condition to that in a plain test piece will have an identical life.

3.6 Deformation Type Failure

Cyclic plasticity may accumulate due to the maximum strain increasing incrementally with cyclic loading, and although failure may occur within the low cycle region, the mechanism causing failure is somewhat different from that of the true LCF failure. Any mechanism involving a uni-directional accumulation of strain, which ultimately results in failure, will be termed 'deformation failure'. This type of failure mechanism does not result in a stable hysteresis loop as shown in figure 3.3 but rather a condition where the mean strain progressively increases monotonically, as suggested in figure 3.10. Failure will eventually be brought about, not due to fatigue, but due either to excessive deformation or exceeding the static load carrying capacity of the structure. The mechanism shown in figure 3.10 is usually referred to as 'ratchetting', and this condition can be obtained either by thermal cycling under a steady load, where non-uniform heating causes expansion to be partially restrained, or by mechanical load cycling due to complete reversal of the external loading not completely reversing the stress.

Strain accumulation under the action of cyclic loading may also occur due to 'unstable material behaviour', sometimes referred to as cyclic creep. This phenomenon has been observed for push—pull and repeated tension tests, where constant amplitude load cycling produced a cumulative strain which increased progressively in a tensile direction during both the push—pull and the repeated

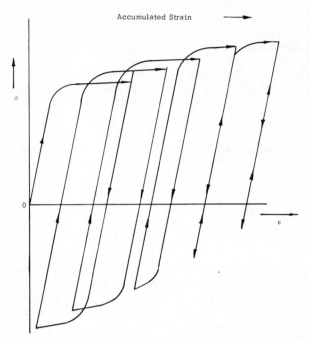

Figure 3.10 *Deformation type failure (ratchetting)*

tension tests[12]. This effect may be represented diagrammatically as shown in figure 3.11, which indicates the accumulated strain due to the ratchetting effect. This phenomenon has also been observed in high strain constant amplitude reversed torsion tests[13], where an axial accumulation of strain is obtained. Further, it has been demonstrated[14] that a constant tensile axial load, superimposed on reversed torsional cycling, amplifies the accumulation of axial strain, and deformation type failure may occur rather than failure by low cycle fatigue.

    In general, if a superimposed static load acts on a system in a different direction from that of the cyclic plastic strains due to dynamic loading then a special

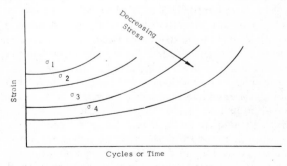

Figure 3.11 *Diagrammatic representation of cumulative strain due to ratchetting (unstable material behaviour)*

deformation type failure may occur, known as 'incremental collapse'. This condition is brought about due to the cyclic accumulation of irrecoverable deformation, caused or aggravated by the action of the superimposed so-called 'follow-up load' on the cyclic strains. Because of the complex nature of incremental collapse, it is not possible to represent diagrammatically the material behaviour. However, for this type of mechanism, it is possible for the cyclic strains in one direction to be presented as in figure 3.3, whilst in some other direction, i.e. the direction of the follow-up load, an accumulation of strain may occur, as demonstrated in figure 3.11. Consequently, under these circumstances failure may occur either by low cycle fatigue (LCF) or incremental collapse, depending upon whether the plastic strain energy or the cumulative strain reaches the critical value first.

3.7 Failure Mechanisms in Practice

It is of interest to consider a possible local material behaviour relationship for an aero-engine turbine disc. Such a disc will generally be a machined forging with an integral shaft or with a flange on to which the shaft may be bolted. The disc has around its perimeter provision for the attachment of the turbine blades[15]. Since a turbine disc will be subjected to a combination of high rotational stresses, thermal stresses, and creep, the material response will be somewhat complex. In particular, it will be of some interest to consider the local behaviour in the vicinity of the blade fixing. A *hypothetical* local material behaviour relationship is shown in figure 3.12 and for simplicity many of the complicating factors are not included. It is assumed that the rotational stresses and thermal stress may be considered separately, although in reality there will, of course, be some interaction. Since there will be a temperature gradient through the disc in a radial direction, differential expansion and contraction will create thermal stresses. At the perimeter, i.e. where the blades are attached, since this will be at a higher temperature than the centre of the discs, the thermal stresses will be compressive. Figure 3.12 indicates

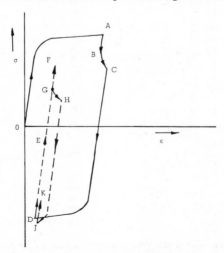

Figure 3.12 *Hypothetical local material behaviour for rim of turbine disc*

diagrammatically the effects of these factors on the local material behaviour. Assuming that start-up occurs cold, and that the disc rotates at high speed, tensile stresses due to rotation will be set up, and if these stresses are in excess of the yield strength of the material, local yielding will occur; this is represented by the path OA. During warm-up, which is assumed to take place whilst the disc is rotating at constant speed, thermal stresses will be set up, and since the rim is hotter than the rest of the disc, these stresses will be compressive, as represented by the path AB. If the disc continues to rotate at high speed at elevated temperature, some stress relaxation may occur, as shown by the path BC. The second half cycle occurs when shut down takes place. The path CD represents the material behaviour when rotational stresses are removed (assumed at constant elevated temperature), and whether yielding in compression is obtained or not depends upon the factors previously discussed.

As the disc cools down, the thermal stresses will be removed, as indicated by the path DE. Thus, at the end of one complete cycle, i.e. on start and stop, the local material behaviour is represented by the path OABCDE. Subsequent cycles may be as illustrated by the dotted path; EF represents reapplication of rotational stresses; FG the reapplication of thermal stresses; GH the possibility of further stress relaxation whilst rotating at elevated temperature, and removal of thermal stresses (cool-down) by JK. Figure 3.13 indicates an alternative hypothetical local material behaviour pattern if yielding occurs on subsequent starts.

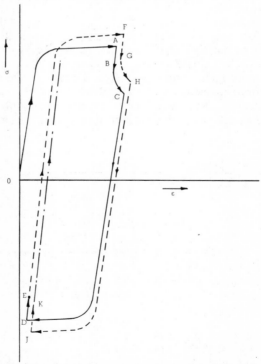

Figure 3.13 *Alternative hypothetical local material behaviour for rim of turbine disc*

It is evident from this discussion that the local material behaviour in the vicinity of the blade fixing (a typical concentration feature) for a turbine disc is far from simple. The actual failure mechanism is obviously complicated, and involves a combined LCF and deformation type failure.

3.8 Concluding Remarks

In this brief chapter, an attempt has been made to present the possible material behaviour relationship which might be expected to be obtained under dynamic loading situations in the low cycle region. If low cycle failures are to be minimised, it is essential that local material response be understood, so that quantitative assessment of integrity can be obtained with a high confidence level. However, the assessment of local material behaviour under dynamic loading, on a quantitative basis, is complex, and predicting such behaviour is probably one of the most important factors still requiring investigation.

From the discussions, it may be concluded that failure in the low cycle region may occur due to either low cycle fatigue or a form of deformation failure, or possibly a combination of both. Thus, it is necessary to be able to determine the accumulation of plastic strain energy or plastic strain under dynamic loading, and also to assess the critical values for the material under consideration. Only then is it possible to assess structural integrity reliably.

The discussion outlined assumes constant amplitude loading, whereas in practical situations it is more likely that a component will be subjected to non-constant amplitude loading. Consequently, if the material behaviour relationship can be predicted for different load levels, it still remains to assess the manner in which cyclic plastic strain energy or strain accumulation builds up to a critical value.

References

1. Esin, A., The Microplastic Strain Energy Criterion Applied to Fatigue, *J. Basic Engrg.*, **90** (1), 28–36 (1968)
2. Jeale, R. Rolls-Royce (1971) Ltd. Private Communication
3. Duggan, T. V., Application of Fatigue Data to Design – Crack Propagation in a Simulated Component Under Cyclic Loading Conditions, *Ph.D. Thesis*, Portsmouth Polytechnic (1973)
4. Manson, S. S., *Thermal Stress and Low Cycle Fatigue*, McGraw-Hill, New York (1966)
5. Rigg, G. J., Investigation into Strain Controlled Fatigue Behaviour in the Range 100–50000 Cycles at Ambient and Elevated Temperatures, *M.Phil. Thesis*, Portsmouth Polytechnic (1975)
6. Miller, K. J., High Strain, Low Endurance Fatigue: A Review, Cambridge University Engineering Department *CUED/C – Mat. TR1* (1969)
7. Manson, S. S., Fatigue: A Complex Subject – Some Simple Approximations, *Expt. Mechanics*, **5** (7), 193–226 (1965)

8. Sandor, B. I., *Fundamentals of Cyclic Stress and Strain*, The University of Wisconsin Press (1972)
9. Morrow, Jo Dean, Internal Friction, Damping and Cyclic Plasticity, *ASTM STP 378*, American Society for Testing and Materials (1965)
10. ASTM, Cyclic Stress—Strain Behaviour — Analysis, Experimental, and Failure Prediction, *ASTM STP 519*, American Society for Testing and Materials (1973)
11. Sachs, G., Gerberich, W. W., Weiss, V. and Lattorre, J. V., Low Cycle Fatigue of Pressure Vessel Materials, *Proc. Am. Soc. for Testing and Matls*, **60**, 512–529 (1960)
12. Tilly, G. P., Cumulative Strain Behaviour of a Nickel—Chromium Alloy and an 11 Per Cent Chromium Martensitic Type of Steel Under the Action of Cyclic Loading, *Applied Mechanics Convention, Proc. Inst. Mech. Engrs*, **180**, Part 31, 403–413, 483–486, London (1966)
13. Chandler, D. C., High Strain Torsional Fatigue, *Ph.D. Thesis*, University of London (1968)
14. Ronay, M., On Second Order Strain Accumulation in Torsion Fatigue, *Tech. Report No. 16*, Institute for the Study of Fatigue and Reliability, Columbia University, New York (1965)
15. *The Jet Engine*, Rolls-Royce, *Ref. T.S.D. 1302*, Derby, 3rd edn (1969)

4

Metallurgical Aspects of Fatigue

4.1 Introduction

There are three main areas in which extensive work on fatigue has been, and continues to be, carried out by metallurgists, usually in collaboration with engineers since fatigue is very much an interdisciplinary problem. These areas are:

(1) Fundamental research into mechanisms of fatigue damage, crack initiation and the different stages of crack propagation, with the objective not only of satisfying scientific curiosity, but also of using this fundamental knowledge in the development of alloys with improved fatigue properties.

(2) Testing of specimens and components to provide property data necessary for design purposes on fatigue crack initiation and propagation lives, and the effects of variation in loading, temperature and environment.

(3) Investigation of service failures in order to diagnose the failure mechanism or combination of mechanisms, to ensure that material quality is satisfactory, in some cases to possibly quantify crack propagation rates and cycles, and in general to provide feedback to the designer from the ultimate (albeit expensive) fatigue test.

As in any area of technology where disciplines overlap and need to interact for mutual benefit there is a problem of communication to be overcome. Therefore in the sections that follow, the metallurgical aspects of fatigue considered are dealt with in fairly general terms and as simply as possible with the minimum of jargon. For a more detailed explanation of this area the reader is referred to Forsyth[1].

4.2 The Reality of Metal Microstructure

Normal structural metals consist of aggregates of crystals commonly called grains, since they very rarely exhibit the external geometric regularity normally associated with crystals. Simply defined, a crystal is regarded as a portion of material in which the atoms or molecules (in the general case) are arranged in a definite pattern characteristic of the solid material at any temperature, known as the crystal lattice.

Since it has a directional structure, the individual crystal is readily shown to possess directional properties. However, in an aggregate of millions of crystals with

65

random orientation this short range directionality would tend to be self cancelling and allow for the approximation made in the assumption of isotropy and homogeneity of material for the simple static design situation.

In this latter situation the designer employs strength values based on either the onset of plastic deformation (yield or proof strength), or instability and fracture (tensile strength) obtained from a tensile test. It is usually assumed that the whole gauge length of the test piece experiences the particular effect uniformly and simultaneously, though this cannot be true on a microscale. The level of these strength values has been improved over the years by variation of the microstructure and therefore the properties of metals by alloying and thermomechanical treatments. A knowledge of the strengthening mechanisms involved has not been necessary for the design engineer to make sensible use of this improved static strength, since the conditions under which the test data are derived bear a reasonable resemblance to the static situations in which they are employed.

In the case of fatigue failure, however, where cracking and fracture can occur at cyclic loads which would be quite safe under static conditions, and where improved tensile strength does not necessarily give an increased fatigue strength (figure 4.1), it is desirable for the designer to obtain a better understanding of the microstructure of the material, with its inherent inhomogeneity and anisotropy.

Reasons for the presence of inhomogeneity and anisotropy in the polycrystalline aggregate include:

(1) Lattice imperfections on an electron-microscopic scale such as: dislocations; vacancies; stacking faults; and lattice strain due to work hardening, solution and precipitation hardening.
(2) The presence of different types of crystals. Most conventional high strength alloys possess a multi-phase structure, such as ferrite and cementite in

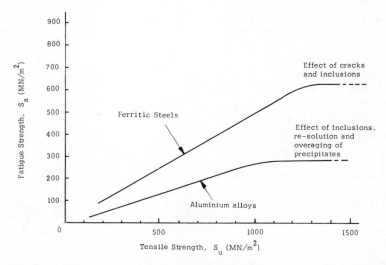

Figure 4.1 *The general trend of correlation between fatigue strength and tensile strength for ferritic steels and aluminium alloys*

ferritic steels, where each phase contributes radically different properties to the resultant strength of the alloy.

(3) Preferred orientation of a significant number of grains. Most high strength components are forgings and acquire directional structures and properties in their working, probably the most extreme example of this is in titanium alloys where slip, the process of plastic deformation, is constrained to very limited crystallographic planes in the close packed hexagonal (CPH) crystal structure.

4.3 Formation and Propagation of Fatigue Cracks

It is generally considered that the process of fatigue can be realistically divided into the following three phases:

(1) a primary stage, i.e. nucleation or crack initiation;
(2) a secondary stage, i.e. crack propagation; and
(3) a final stage, i.e. failure by fracture or some other limiting factor.

These stages are indicated diagrammatically in figure 4.2.

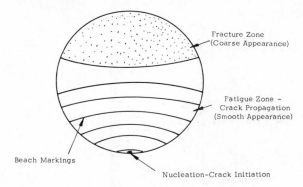

Figure 4.2 *Phases of fatigue failure*

Crack Initiation

Most of the earlier work, prior to Thompson and Wadsworth's review[2] in 1958, tended to concentrate on the crack initiation phase, since the crack propagation phase was erroneously considered to be a minor part of fatigue life. The subsequent review by Plumbridge and Ryder[3] in 1969, summarises comprehensively the present 'state of the art' on crack initiation.

The initiation of a fatigue crack is usually at the metal surface since at that location:

(1) the stress concentration is greatest, and in practice in engineering components initiation is usually at a stress concentration feature such as a change in section, an undercut or inadequate radius, etc.;

(2) the surface crystals are less mutually supported than the interior crystals and thus slip more readily;
(3) atmospheric action is effective.

As might be expected there have been exceptions to the general rule of surface initiation, mainly in some specific service failures where fatigue cracks have unambiguously initiated inside the specimen. In these cases it seems likely that some weak microstructural interface acted as a nucleation site.

In the absence of stress-raisers, in smooth surfaces of ductile metals, initiation involves the build-up of permanent fatigue damage, i.e. permanent microstructural and surface topography changes, followed by crack formation in this disturbed area of the surface. The sequence of events has been usually found to be:

(1) slip band formation;
(2) the formation of extrusions and intrusions;
(3) crack development at intrusions as shown diagrammatically in figure 4.3.

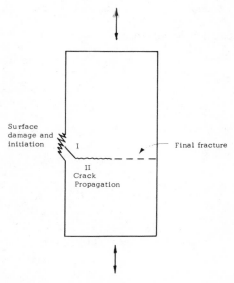

Figure 4.3 *Schematic diagram of the different stages in the fatigue failure process*

Slip bands represent the planes on which the component parts of crystals have suffered relative sliding, i.e. plastic deformation, and occur on the smooth surface of a metal during both monotonic and cyclic loading.

As shown in figures 4.4 and 4.5 the formation of extrusions and intrusions appears to be an extension of the slip band formation process, requiring slip on intersecting planes and some irreversibility of this process. Both Cottrell and Hull[4] and Forsyth[5] have proposed mechanisms for extrusion/intrusion formation involving slip reversal and dislocation climb by association with vacancies or excess atoms.

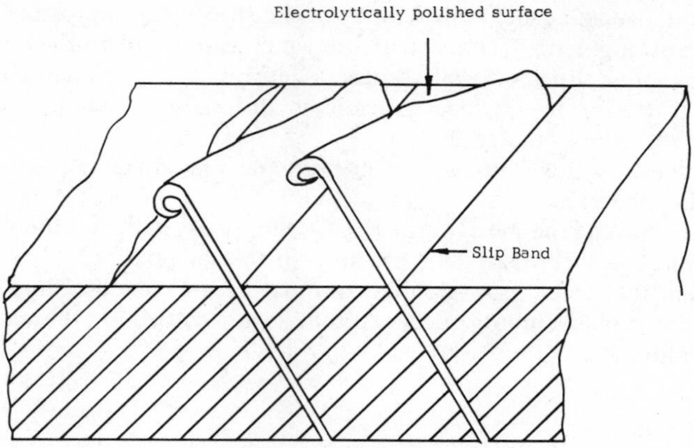

Electrolytically polished surface

Slip Band

Figure 4.4 *Slip-band extrusions*

It is almost impossible to define exactly when a microcrack develops from an intrusion and Stage I crack propagation begins, and in terms of engineering design the question is clearly an academic one.

The practical results of the enormous amount of metallurgical research carried out into fatigue crack initiation are considered to be as follows:

(1) The fundamental mechanism of initiation is well understood.
(2) The microstructural requirements for maximising resistance to initiation have been established, namely a structure which resists precipitate

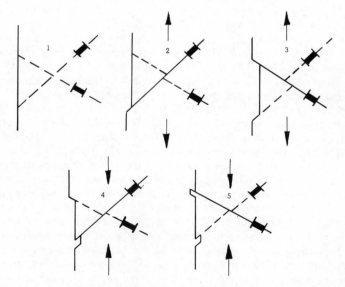

Figure 4.5 *Cottrell and Hull's model for producing extrusions and intrusions*

dissolution, overaging and annealing by cyclic stressing — suggesting dispersion strengthening, and a structure which is resistant to the occurrence of multiple cross slip and localised plastic deformation on planes which intersect a free surface (giving extrusions/intrusions) — which dictates a low stacking fault energy material.

Unfortunately this formula does not provide a practical or economic engineering material.

(3) The significance of the metal surface, particularly in HCF, and the value in some cases of surface treatments has been further emphasised.

(4) It is evident that crack propagation is of much greater practical importance since even in a plain smooth specimen this stage may occupy the vast bulk of fatigue life.

Crack Propagation

It is clear that the fatigue performance of most components and structures is more dependent on the ability of cyclic stress to propagate rather than initiate cracks. This fact is evident since microcracks are known to appear very early in the lives of specimens and components and notched high strength material may have propagating cracks effectively throughout service life.

Forsyth[6] found that in ductile metals fatigue cracks propagated in two distinct stages, designated Stage I and Stage II crack growth.

Stage I growth, sometimes known as slip-band growth, entails the deepening of the initial crack (intrusion) on a plane or conjugate planes of high shear stress. Figure 4.6 shows this as initially observed by Forsyth. This mode of crack growth may be totally absent in a sharply notched highly stressed situation, yet it may account for up to 90% of the life of smooth ductile specimens at low stresses.

During Stage II growth, the general plane of the crack is normal to the direction of the maximum tensile stress (figure 4.6). The Stage I to Stage II transition is generally supposed to result from the reduction in the ratio of shear to direct stress at the crack tip as it moves from the weakly constrained surface crystals into the interior where mutual crystal reinforcement inhibits slip. The location of this transition often occurs at some microstructural discontinuity in the material; however, many exceptions have been reported. The criterion for growth in Stage II is primarily the value of the stress intensity at the crack tip, discussed in Chapter 6.

The fracture surfaces produced during Stage II growth usually show 'striations' or 'ripples', especially in very ductile metals such as aluminium and austenitic stainless steel. This effect was first observed by Zappfe and Worden[7] and Stubbington and Forsyth[8] using the optical microscope at magnifications of only x1500, as shown in figure 4.7. Subsequent work using electron microscopy on programme loaded specimens indicated that each striation was produced by a single cycle[9,10]. The converse statement that every stress cycle produces one striation is not necessarily valid, since the crack tip stress resulting from some cycles may be too small to produce an increment of crack length, especially in a random loading situation. This fracture phenomenon provides the basis for attempts at quantifying crack propagation, by striation counting (thereby estimating spacing) giving a direct measure of crack propagation rate[11]. Striations are not to be confused with 'beach markings', the characteristic fatigue pattern often visible with the naked eye on

Figure 4.6 *Series of Stage I cracks with one in transition to Stage II propagation in pure aluminium (Forsyth)*

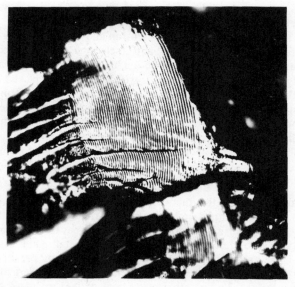

Figure 4.7 *Characteristic striations or ripple marks on a Stage II fatigue fracture surface of an Al-Zn-Mg alloy. Each ripple represents a position of the advancing crack front and each was produced by a single load cycle (Stubbington and Forsyth)*

fatigue fractures (figures 4.2 and 4.21). This latter effect is due to pauses during propagation and therefore variation in oxidation of the fracture surface.

The factors governing the extent of crack growth per cycle and why the growth during one cycle is limited, are not fully understood and are the subject of fairly extensive current research. Attempts to explain these phenomena require some understanding of fracture mechanisms.

4.4 Fracture Mechanisms

This section contains a simple account of the mechanisms involved in ductile, cleavage and fatigue crack propagation. For a more comprehensive explanation of the whole field of fracture mechanisms, in still relatively simple terms, the reader should consult Petch[12].

A piece of structural metal is a polycrystalline aggregate, with crystalline regions separated by disordered regions giving crystal boundaries. In the simplest terms there are two possible ways in which the material can fracture, either through the crystals (trans- or intracrystalline) or between the crystals showing a preferred path for the boundaries (intercrystalline).

In the case of trans-crystalline fracture under monotonic load there are then two main possibilities, either cleavage or ductile fracture. Considered very simply on the atomic scale, cleavage involves the direct pulling of atoms apart on crystal planes normal to tensile stress whilst ductile fracture involves sliding atoms over one another under the action of shear stress. In practice pulling of atoms apart is drastic and requires about $E/5$, whereas shearing atoms over one another is less drastic and even in the absence of dislocations only requires about $G/10$, where E is Young's modulus and G the shear modulus. Thus stress will normally be relieved by plastic deformation before the rupture of atomic bonds can occur.

Since E is a measure of resistance to extension and G of resistance to shear, a low G/E ratio favours ductile behaviour, and since $G = E/2(1 + \nu)$ where ν = Poisson's ratio) low G/E is obtained with a high Poisson's ratio material[12]. With high ν values (i.e. face centred cubic metals with $\nu = 0.3-0.4$) plastic deformation occurs before there is any possibility of tensile rupture of atomic bonds. On the other hand body centred cubic (BCC) metals (e.g. α iron, as in ferritic steels, with $\nu = 0.28$) can be ductile or cleave in a brittle manner, although even in the latter case some small amount of prior plastic deformation always occurs. Figure 4.8 shows diagrammatically the general form of ductile and cleavage fractures.

Ductile Fracture

Normally metals fracture by the ductile mechanism, a relatively high energy absorption process, and completely pure inclusion-free material may fracture simply by plastic deformation to 100% reduction of area. However, in more practical situations second phase particles, often non-metallic impurity inclusions are important, since microscopic cavities are formed either by breaking of the particles under the plastic deformation of the matrix or by rupture at the particle/matrix interface. Under deformation the cavities ('micro-voids') grow together ('coalesce') by a process of internal necking and so produce a ductile fracture crack.

Thus the form and number of second-phase particles is important in ductile fracture, and a low strain hardening rate favours the early onset of cavity coalescence, often referred to as 'micro-void coalescence'. Where the plastic zone at

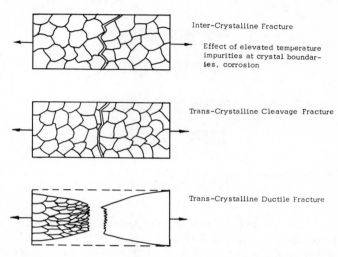

Inter-Crystalline Fracture

Effect of elevated temperature impurities at crystal boundaries, corrosion

Trans-Crystalline Cleavage Fracture

Trans-Crystalline Ductile Fracture

Figure 4.8 *Diagrammatic illustration of simple intercrystalline and trans-crystalline cleavage and ductile fracture modes*

the crack tip (see Chapter 6) is small compared to the through thickness of the metal, the crack propagates as a flat plain strain fracture, but with larger plastic zones it propagates as an inclined shear fracture.

Cleavage Fracture

Cleavage fracture, which is common in BCC at low temperatures and CPH metals, is a relatively low energy absorption process and occurs at very low plastic strains as compared with ductile fracture. First there is the formation of a crack nucleus for example by the fracture of a brittle second phase, followed by a rapid propagation phase. Cleavage is favoured by factors that raise the stress available for propagation of the nucleus, namely triaxial tensile stress, decrease in temperature, and hardening mechanisms. Finer grain size whilst it raises the yield stress (with corresponding increase in hardness), reduces the size of the initial nucleus therefore fine grains are important inhibitors of cleavage.

Fatigue Crack Propagation

This is normally a transgranular mode of fracture, although the effects of elevated temperature, corrosive environment and brittle grain boundary phases, may modify this general rule. The basic mechanism of fatigue crack propagation is that of striation formation with variations on this basic mechanism dependent on microstructure, ductility, stress intensity range (see Chapter 6) and environment. A

variety of surface profiles of fatigue-fracture striations have been reported by
several workers and these have been summarised by Laird[13], as shown in figure 4.9.

Forsyth et al.[14] have distinguished between two sorts of striations: 'ductile'
(type A) and 'brittle' (type B). 'Ductile' striations consist of alternate light and dark

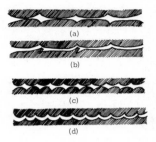

Figure 4.9 *Diagrammatic representations of reported profiles of fatigue-fracture striations
according to several workers (Laird)*

bands lying on irregular non-crystallographic plateaux, whilst 'brittle' striations lie
on crystallographic facets (figure 4.10).

There is some dispute as to the precise mechanism of striation formation,
although it is generally accepted that it involves alternate blunting and
re-sharpening of the crack tip. The mechanisms of plastic blunting proposed by
Laird[13] and McMillan and Pelloux[15] are illustrated diagrammatically in figure 4.11,
and both would appear reasonable for ductile striation formation. Whether crack
propagation proceeds by either purely 'ductile' or 'brittle' striation formation or a
combination of both will depend on the material, temperature and environment,
and may to some extent control the rate of propagation for a particular stress
intensity range (see Chapter 6).

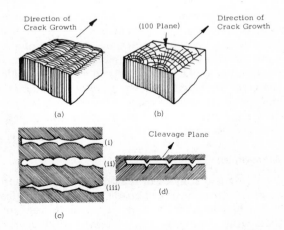

Figure 4.10 *Different types of ductile and brittle fatigue striations (McMillan and Pelloux)*

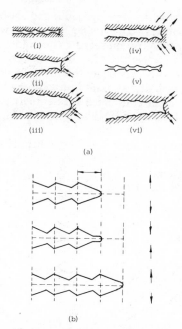

Figure 4.11 *Diagrammatic representations of the mechanism of striation formation by plastic blunting: (a) according to Laird; and (b) according to McMillan and Pelloux*

In addition to the striation formation type of crack growth, relatively large scale steps of crack advance may occur by ductile, microcleavage or intergranular mechanisms. Richards and Lindley[16] report these effects for a wide range of steels, as does Ritchie[17] for a low alloy steel and a high nitrogen mild steel. These larger scale crack increments (cf. striations) are known as monotonic or static modes of crack propagation. It is clear that these departures from the simple striation formation growth mechanism cause increased crack propagation rate. The tendency for the occurrence of such static modes appears to be increased by an embrittling condition in the material which causes either microcleavage, or intergranular modes[16,17], and by higher values of maximum stress intensity approaching the net section yield stress which causes patches of microvoid coalescence (ductile static mode) in more ductile materials[16]. An exception to these latter general rules was the observation of intergranular static modes in fatigue crack propagation in tempered martensitic structures at low stress intensity range. However, it is argued[16] that this mechanism is environmentally controlled.

4.5 The Fatigue Limit

Within the investigated life limit of approximately 10^{10} cycles, the S–N curve for most metals shows a continuing fall with decreasing slope. This observation is compatible with the dependence of crack initiation on a slip process (the movement of dislocations) in microscopically heterogeneous and anisotropic material, i.e. some dislocation movement will be inevitable, however low the cyclic stressing.

Ferritic steels, however (and certain specific aluminium, magnesium and titanium based alloys), show a tendency for the S—N curve to become asymptotic (a 'knee' is observed in the S—N curve) and the stress amplitude level at which this occurs is termed the fatigue limit; this limit is usually reached at about $10^5 - 10^7$ cycles. The phenomenon can be considered as essentially a threshold stress amplitude below which a fatigue crack will not initiate. Figure 4.12 indicates the general behaviour of steel and other alloys.

Non-ferrous metals in general, do not exhibit a distinct fatigue limit, although there is a gradual levelling out of the S—N curve. The term endurance limit is often used for the non-ferrous metals, where this is the stress amplitude for a defined life, e.g. $N = 10^8$, or 5×10^8 cycles.

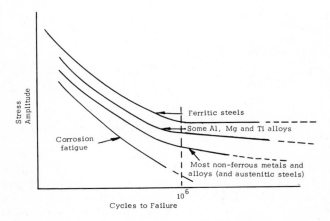

Figure 4.12 *The general trend of S—N behaviour*

It is notable that the removal of carbon and nitrogen (both interstitial alloying elements) from mild steel tends to eliminate both the fatigue limit and the yield point[18]. Figure 4.13 shows the loss of the fatigue limit with the removal of carbon alone from a mild steel. Thus the fatigue limit is usually attributed to the dislocation locking effects of strain aging. Strain aging is the process by means of which the yield point of mild steel can be raised by static overstrain followed by an aging period. The 'weak spots' (dislocations) which initially permit yielding at low stresses are subsequently 'sealed' (dislocation locking) and strengthened by the migration of the carbon atoms during the aging period.

A possible relationship exists between the fatigue limit and the yield point. Cyclic stresses above the fatigue limit value readily unlock dislocations from their interstitial sealed positions so that if a mild steel specimen is statically tensile tested after a period of cycling above the fatigue limit, the static yield point is less marked.

It is commonly believed that the extension of fatigue life by preliminary cycling at sub-service stresses, known as 'coaxing'[1] is also attributable to strain aging, although the evidence is not really conclusive.

Clearly the knowledge of a fatigue limit for a material or an endurance limit defined beyond designed life cycles is a useful asset to the designer provided the conditions under which the data were obtained are truly relatable to the service conditions.

In many ways analogous to the fatigue limit, which as mentioned earlier may be considered as a threshold stress amplitude for fatigue crack initiation, is the stress

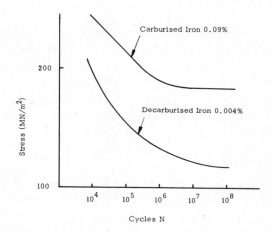

Figure 4.13 *S—N curves for carburised and decarburised iron (reproduced from Smallman*[18] *by permission)*

intensity range threshold for crack propagation. This is the value of stress intensity range below which an existing crack in a metal will not propagate (at least in practical terms), and where this condition exists it is of great practical value to the designer. The threshold condition for non-crack propagation will be considered in Chapter 6.

4.6 Surface Condition and Treatment

Surface Condition

Fatigue strength, as indicated by the fatigue or endurance limit, i.e. essentially resistance to crack initiation, is very dependent upon the weakest most highly stressed crystals; these usually lie at the metal surface (see sections 4.2 and 4.3), although there are exceptions to this general rule. Unless special counter-measures are employed, these surface crystals represent the weakest regions because of the absence of neighbouring grain support on the free surface. These crystals also experience the stress concentrating effects of sectional change, surface roughness and nominal-stress gradients.

The fact that in most cases of fatigue failure cracking commences at the surface further confirms the design importance of surface condition. 'Surface condition' may involve all the following considerations.

(1) Surface Geometry

The presence of acute notches is generally harmful but all material conditions do not have the same sensitivity to a given notch form.

(2) Residual Stress

The inducement of residual compressive stress may prove beneficial in reducing the effective stress acting at the surface whereas residual tensile stress is harmful.

(3) Strain Hardening

Local surface strain hardening is usually beneficial by increasing strength and therefore slip resistance, but cyclic strain softening in some materials may to some extent offset this effect.

(4) Structural and Compositional Variations

Nitriding, carburising and flame hardening are generally beneficial, whilst decarburisation and the selective removal of alloying elements are usually harmful.

(5) Corrosion Sensitivity

The joint action of corrosion and fatigue loading is particularly damaging, and the sensitivity of a metal to a particular environment is influenced by the physical condition of its surface. The effects of environment are discussed in more detail in section 4.7.

Ideally the understanding of these effects requires their separate investigation, however, with the exception of surface geometry the required isolation is virtually impossible. For example, strain hardening is always accompanied by microscale residual stress and both notch sensitivity and corrosion behaviour are affected.

The importance of residual stress depends upon:

(1) its magnitude, sense and direction relative to the externally applied stress;
(2) the material properties; and
(3) its stability with respect to the time—temperature recovery processes occurring during cyclic stressing.

Residual stress may be regarded as a superimposed mean stress in that it locally modifies the overall loading. Additionally, its introduction can modify the material's response to that loading. In high stress cycling, the correspondingly high plastic strains must modify any pre-existing residual-stress distribution very early in a component's life and it is thus very doubtful if residual stress considerations are important for failure lives of less than about 10^4 cycles.

Surface Treatments

(1) Strain Hardening and Residual Stress

Some industrial processes such as machining, cold-forming, heat treatment, etc., introduce residual stress and strain hardening as a side effect of their main purpose. Other treatments are specifically applied to improve fatigue strength by appropriately influencing these factors. Such treatments include skin rolling, mechanical polishing, shot peening, flame hardening, case carburising and nitriding.

An up-to-date review of the effects of surface treatments has been made by Frost et al.[19] The effectiveness of a treatment is judged on its overall effect, so that processes inducing compressive residual stress, stable strain hardening, increased stable strength by heat treatment and a surface free from stress-concentrating features are generally beneficial. Thus shot peening and skin rolling eliminate the stress-concentrating marks left by grinding whilst introducing a favourable residual stress system together with strain hardening. It is reported that shot peening induces a biaxial compressive stress of half the yield stress or more in a surface layer in the region of 0.1–0.5 mm deep[20].

Contact pressure is more easily controlled in the surface-rolling process than in shot peening, and can attain higher values, therefore a deeper surface layer of material can be worked by rolling than by peening. Examples of increases in the fatigue limit of shafts due to rolling are given by Almen and Black[21]. Rolling is particularly effective in increasing the fatigue limit of notched specimens or components, e.g. cold rolling of the journal and web fillets of crankshafts[22]. Frost et al.[19] strongly recommend that whenever the possibility of initiation of fatigue cracks at discontinuities or notches is anticipated, then surface-working processes should be employed to counteract this, especially surface rolling.

(2) Surface Hardening

There are three main metallurgical processes commonly used to produce a hardened layer on the surface of steel components, namely, induction or flame hardening, carburising and nitriding. All of these processes harden the surface and produce metallurgical structure volume changes which induce high compressive residual stresses in the surface layer. The greatest improvement in fatigue limit is observed where stress concentration features are present in components[19].

(3) Metal Platings

Chromium may be electroplated on to steel components either as a wear resistance coating or to build up a worn or under-sized region; also nickel and cadmium are often used for corrosion resistance on steel. The effect of such coatings on the fatigue strength of the substrate steel has been reviewed by Hammond and Williams[23]. As-deposited chromium contains internal cracks and tensile residual stress whose magnitude depends on the plating conditions, and on the thickness of the deposit. In general it would seem that, in the case of those platings which make an interatomic bond with the steel substrate, the fatigue strength of the plated component depends on the discontinuities and residual stresses developed during the plating process. Because the crack growth characteristics of steel tend to be similar once the stress level exceeds that necessary to cause a crack in the plating to grow, the strength of the steel substrate will make little difference to the fatigue limit of a plated specimen. However, any method, such as inducing high compressive residual stresses in the surface layers of the substrate, that prevents the growth of these inherent cracks will improve the fatigue strength of the plated specimen[19].

(4) Anodising

This process of inducing an oxide layer on the surface of aluminium alloys, for improved wear and corrosion resistance, in general reduces fatigue strength to some extent dependent on the specific process variables. Sealing of the coating appears to minimise this reduction in fatigue strength. Again the effects of this particular surface treatment are considered in some detail by Frost et al.[19]

4.7 Environmental Effects

Under a combination of static tensile stress and a corrosive environment some alloys form cracks and fail by 'stress-corrosion'. Well known examples of this phenomenon are the season cracking of cold worked brass, the caustic embrittlement of riveted mild steel boiler seams and chloride stress-corrosion of austenitic stainless steel. The general corrosion resistance of the alloy is not normally impaired, but a susceptibility to cracking in specific environments is observed. Very often the stress component is residual stress as the result of cold working, and in many cases this can be obviated by stress-relief annealing. Whilst stress-corrosion is a serious engineering problem it is confined to specific alloy/environment situations, but the effect of corrosion on fatigue behaviour is far more general.

Except at very short lives the combination of any corrosive environment and fluctuating stress results in reduced fatigue strength and the disappearance of the fatigue limit (figure 4.12) and this effect has become known as 'corrosion-fatigue'. Several reviews of the effect of environment on fatigue strength have been carried out[24-27].

In general the conjoint effect of the cyclic stress and corrosive components of the failure mechanism is greater than the sum of the separate effects of each contributor. In corrosion fatigue, microcracks develop at lower stresses than in-air tests, and often many surface cracks develop and grow to significant sizes before one reaches such a size that failure occurs.

The effect of a corrosive medium on fatigue strength is most severe when it is applied while the specimen is being cyclically stressed and where oxygen (in the atmosphere) has free access to the metal surface. Thus if the corrodent is sprayed or dripped onto the metal surface, the subsequent fatigue strength is lower than if the specimen is immersed completely. Experiments on plain specimens of mild steel, showed marked reductions in fatigue strength in sprayed water and 3% salt solution, whereas complete immersion in de-aerated water and salt solution showed no reduction in fatigue strength on the in-air properties (figure 4.14)[28].

In general the corrosion-fatigue strengths of plain specimens tend to follow the pattern of the corrosion resistance of the materials concerned, e.g. the corrosion-fatigue resistance of 18/8 stainless steel is usually better than that of a low-alloy steel of similar tensile strength. Thus austenitic steel together with Monel alloy, aluminium, phosphor, and beryllium-bronzes, all display relatively high resistance to corrosion fatigue. Because corrosion-fatigue strength is sensitive to small variations in the corrosive media and small changes in temperature and speed of testing, the corrosion-fatigue strengths obtained in the laboratory should only be used as indicative of a general pattern of behaviour. Furthermore, service components usually experience the corrosive environment for vastly longer times than laboratory specimens, even though the number of stress cycles applied may be similar.

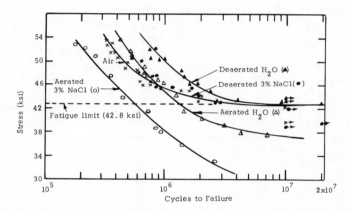

Figure 4.14 *Effect of air and aerated and de-aerated distilled water 3% NaCl solution on fatigue behaviour of 1015 (A) steel at 25°C (Duquette and Uhlig)*

The use of metal coatings is well known in corrosion protection, but they may actually shorten corrosion fatigue life for the following reasons:

(1) pure metal coatings are mechanically weak and readily acquire fatigue cracks which they can transmit to the underlying metal;
(2) electrodeposits may be in a state of residual tensile stress;
(3) dip-coatings often include brittle alloy layers;
(4) a discontinuous coating could act cathodically relative to the metal requiring protection and thereby accelerate corrosion, e.g. chromium and nickel on steel (although zinc and cadmium coatings are anodic to a steel substrate and are advantageous[19]).

Unless attempted by applying appropriate and adherent paint/oil coatings or nitriding in the case of special steels, it is usually better to manufacture a component from a metal inherently possessing adequate corrosion-fatigue resistance rather than attempt to confer resistance as a surface property alone. An extensive survey of corrosion-fatigue data is presented by Cazaud[29,30] referring to different materials, corrosive solutions and types of stressing.

4.8 Fretting Fatigue

The surface damage which occurs when two well-mating surfaces experience slight relative motion is known as fretting or fretting corrosion, and the most up to date review is by Waterhouse[31].

Fretting corrosion is both material and atmosphere dependent. Oxygen generally, and nitrogen specifically in the case of steel, are harmful, but in contrast to ordinary corrosion, a rise in temperature and humidity (possibly acting as a lubricant) are beneficial.

The mechanism of fretting corrosion involves a combination of the following:

(1) Surface contact at high spots resulting initially in plastic flow and cold welding with subsequent weld rupture, the liberation of metal particles and possible local attainment of melting temperature.

(2) Unless oxygen is excluded, the metal particles become oxidised and hence more abrasive. This is particularly the case with aluminium oxide which appears in a crystalline form whose normal formation temperature is in excess of $1000°C$.

(3) The exposure of bare active metal when protective oxide films are ruptured. Particularly at lives above 10^6 cycles, fretting corrosion sites can be the source of fatigue cracks, and the situation is aggravated by the localised pitting and sectional change in the component.

Investigations indicate that fretting corrosion is more severe between soft metals but that the fatigue coupled effect is worst between hard metals. With dissimilar metals, both are affected in spite of hardness differences.

Countermeasures to fretting corrosion and hence to fretting fatigue include:

(1) The prevention of relative motion, e.g. by raising the contact pressure or increasing the friction coefficients by electrodeposition of certain metals.

(2) Alternatively a considerable reduction of the friction coefficient may be beneficial, e.g. inserts of PTFE between leaf springs.

(3) Increasing resistance to abrasion by case hardening, nitriding, shot peening, skin rolling and the formation of lubricant impregnated phosphate or anodised coatings.

(4) The exclusion of the atmosphere, e.g. by flooding with lubricant of low oxygen solubility.

4.9 Elevated Temperature Fatigue

Ignoring any particular heat treatment effects, a rise in temperature generally lowers both the fatigue and static-tensile strength of metals, and the reverse occurs if the test is conducted at lower temperatures.

Mild steel shows maximum fatigue, yield, and tensile strengths in the range $200-400°C$, and this is generally attributed to strain aging.

At higher temperatures plastic flow tends to become predominantly a grain boundary sliding mechanism so that creep rather than fatigue failure becomes increasingly probable (figure 4.15).

Except for conditions when creep and/or corrosion are significant, cyclic speed variation in the normal range of 1–100 Hz has little effect on fatigue strength. Under creep conditions, however, it is found that for a given number of cycles to fracture, the stress decreases with frequency and this effect may be correlated with the influence of temperature and strain rate upon the static tensile strength as follows:

(1) the temperature at which static tensile strength reaches a maximum value depends upon strain rate; and
(2) the temperature at which fatigue strength is greatest depends upon cyclic frequency.

Both grain boundary cracking due to the application of tensile mean stress and corrosive attack are time dependent and hence cause this latter frequency dependence of elevated temperature fatigue strength. Frequency dependence

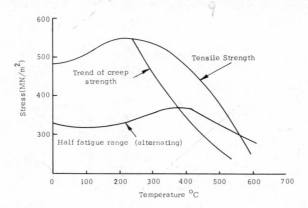

Figure 4.15 *Variation with temperature of tensile strength, fatigue strength and creep strength of a low-carbon steel (Tweedale[32])*

implies a wave form dependence, since the slower the load is applied then the longer is the time per cycle for deformation to occur. Thus where cycles include dwell times at peak stress (i.e. flat peak wave forms), this will cause a decrease in fatigue strength because the higher the ratio of the root mean square stress value to peak stress, the greater is the contribution of corrosion attack and/or grain boundary cracking[19].

A good high-temperature fatigue strength would be expected from a material able to retain a high resistance to cyclic slip within a grain (i.e. initiation resistance), to sliding at grain boundaries (i.e. creep resistance), and to atmospheric attack. Because high creep resistance materials possess these properties, they usually have correspondingly good high-temperature fatigue strength[19].

In many practical engineering situations the factors fatigue, creep and corrosion may be acting simultaneously, as for example in power generation plant, gas turbines and chemical plant. Clearly a very complex problem of predicting life arises with this interaction and this is the subject of much current research and development work. For accounts of some of the most recent work the reader is referred to the proceedings of a recent international conference[33], and for the most up to date design practice recommendations, to the up-dated ASME code for Pressure Vessels[34].

4.10 Fatigue Failures

A fatigue fracture generally consists of two distinct areas, namely a fatigue zone and a final fracture zone. The fatigue zone is due to crack propagation, and because of the continued opening and closing of the crack, this zone has a smooth appearance. The final fracture zone, on the other hand, is due to sudden fracture and has a coarse appearance. A low overstress will result in a high ratio of fatigue zone to final fracture zone, whilst a high overstress is indicated by a low ratio of fatigue zone to final fracture zone; figure 4.16 indicates this diagrammatically for a circular test specimen. Failure usually commences at the outer surface, mainly

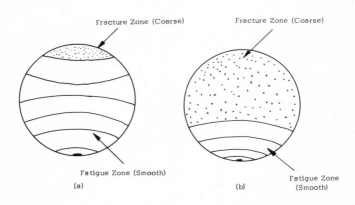

Figure 4.16 *Fatigue fractures. (a) Low overstress; (b) high overstress*

because the maximum stress or strain amplitude occurs there (either because of a stress or strain gradient due to the type of loading, or due to a stress concentration arising from the geometrical configuration), but also because at the outer surface there are unsupported grains which introduce local weaknesses. The shape of the final fracture zone (sometimes referred to as the instantaneous rupture zone) depends upon the nature of the load and the geometry of the parts, and it may also be influenced by such factors as surface hardening, in which case it is possible for fatigue failure to be initiated from beneath the surface. Some typical fracture appearances for different types of loading and degrees of stress concentration are indicated diagrammatically in figures 4.17 to 4.20. Figures 4.21 to 4.27 indicate

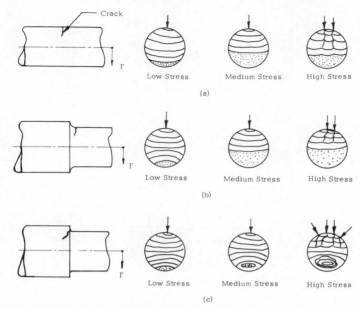

Figure 4.17 *Fluctuating bending fatigue fractures. (a) Zero stress concentration; (b) medium stress concentration; (c) high stress concentration*

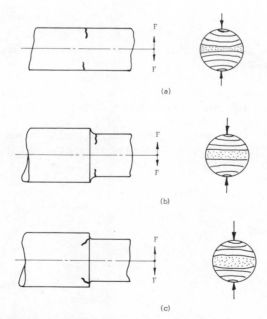

Figure 4.18 *Alternating bending fatigue fractures. (a) Zero stress concentration; (b) medium stress concentration; (c) high stress concentration*

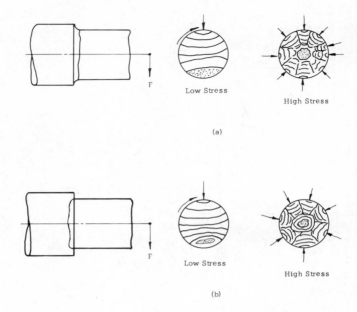

Figure 4.19 *Rotating bending fatigue fractures. (a) Medium stress concentration; (b) high stress concentration*

some service fatigue failures, and illustrate how such failures are frequently influenced by external factors.

Figure 4.21 indicates a bolt stem which failed due to offset load under the bolt head; a pattern of service life is evident from the bands (sometimes referred to as beach markings) of the fatigue zone. The high ratio of fatigue zone to final fracture zone indicates only a small overstress.

Figure 4.22 shows a spur gear with case hardened teeth damaged on the pressure line due to overload; the subsequent fatigue area reflects the metallurgical condition of the diffused carburised case.

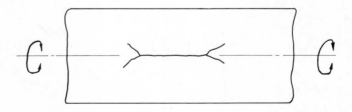

Figure 4.20 *Reversed torsion fatigue fracture*

Figure 4.21 *Stem of bolt failed due to offset load under the head. Pattern of service life is evident from bands of fatigue area*

Figure 4.22 *Case hardened gear teeth having teeth damaged on pressure line due to overload. Subsequent fatigue area reflects the metallurgical condition of the diffused carburised case*

Figure 4.23 *Grinding abuse on thin section together with stress concentration due to cutting away flange at point close to location hole contributed to this failure*

Grinding abuse, together with stress concentration due to cutting away of the flange close to a location hole, contributed to failure of the thin section shown in figure 4.23.

Failure originating in the fir tree fixing of a gas turbine rotor, due to operating at near resonant conditions, is shown in figure 4.24; the service programme is depicted by the fatigue propagation pattern.

Figure 4.25 indicates how a fatigue origin due to frettage occurred in a bevel gear. Flexing of the web produced frettage under the bolt head which acted as a fatigue nucleation area. The remedy for preventing further failures of this kind included improved torque tightening of the bolts.

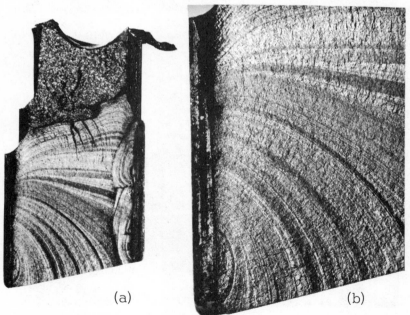

(a)   (b)

Figure 4.24 *Service programmes depicted in fatigue crack propagation pattern. Failure originating in fir tree fixing of gas turbine rotor operating at near resonant condition. (a) General view; (b) nucleation region*

(a)

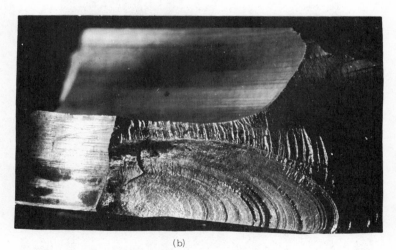

(b)

Figure 4.25 *Frettage as a fatigue origin in a bevel gear. (a) General view showing origin and subsequent crack propagation; (b) view of fatigue zone surface*

Two views of a splined shaft, nitrided for improved wearing qualities, are shown in figure 4.26. The nitrided case has cracked due to torsional overload, and acts as a fatigue origin. Subsequent crack propagation is typical of a fine grain steel.

A high strain low cycle fatigue failure of a chromium—vanadium steel torsion bar is shown in figure 4.27. That this is a high strain failure is evidenced by the uniformity of the four separate origins influenced by the flange bolt holes.

Many more examples of typical fatigue failures have been recorded and discussed

(a)

(b)

Figure 4.26　*Two views illustrating fatigue failure of nitrided splined shaft. Torsional overload cracked nitrided case which then acted as fatigue origin*

(a)

(b)

Figure 4.27 *High strain LCF failure of a chromium—vanadium—steel torsion bar. (a) General view of torsion bar; (b) close-up view showing uniformity of four separate origins*

in the literature[35-38], but the reader is especially referred to the recently published Metals Handbook[39] which contains a comprehensive treatment of the engineering and metallurgical aspects of fatigue failure.

4.11 Concluding Remarks

The ability to easily plastically deform is a major asset of metallic materials, since it is the basis of toughness and fabricability. This same property, is, however, the basic reason for metal fatigue — therefore this failure mode is a fact of material life which must be allowed for, staved off, and quantified as far as possible, but is not likely to be totally eliminated.

Although the mechanism of fatigue crack initiation (an important factor in HCF strength) is now well understood, the development of intrinsically resistant engineering materials does not appear promising, and practical measures to resist initiation appear to be confined to surface control and treatment.

It has become evident in recent years that very often crack propagation is of much greater practical importance than initiation, since even in plain smooth specimens this stage may occupy the greater proportion of LCF life and the bulk of life in notched situations. Even more important is the fact that most engineering structures and components contain defects introduced either during manufacture or during service from sources other than fatigue, and these defects will normally propagate under cyclic loading. The acceptance of this latter philosophy has led to the application of fracture mechanics principles to fatigue crack propagation.

References

1. Forsyth, P. J. E., *The Physical Basis of Metal Fatigue*, Blackie (1969)
2. Thompson, N. and N. J. Wadsworth, *Advances in Physics*, Vol. 7, 72 (1958)
3. Plumbridge, W. J. and D. A. Ryder, The Metallography of Fatigue, *Met. Rev.*, **136**, The Metals and Metallurgy Trust (1969)
4. Cottrell, A. H. and D. Hull, *Proc. Roy. Soc.*, **A242**, 211 (1957)
5. Forsyth, P. J. E., A.S.T.M. Publication No. 327, p. 21 (1959)
6. Forsyth, P. J. E., A two stage process of fatigue crack growth, *Symposium on Crack Propagation*, Cranfield, England (1961)
7. Zappfe, C. A. and C. D. Worden, *Trans. Amer. Soc. Metals*, **41**, 396 (1949)
8. Stubbington, C. A. and P. J. E. Forsyth, *R.A.E. Tech. Note* (Met/Phys. 334) (1962)
9. Forsyth, P. J. E. and D. A. Ryder, *Aircraft Eng.*, **32**, 96 (1960)
10. McMillan, J. C. and R. M. N. Pelloux, Fatigue Crack Propagation, *A.S.T.M. Special Tech. Pub.* No. 415, p. 139 (1967)
11. Jacoby, G., Application of Microfractography to the Study of Crack Propagation under Fatigue Stress, NATO-AGARD Report of 541 (1966)
12. Petch, N. J., The Practical Implications of Fracture Mechanics, *Introductory Lecture*, Inst. of Metallurgists Spring Meeting (1973)
13. Laird, C., Fatigue Crack Propagation, *A.S.T.M. Special Tech. Pub.* No. 415, p. 139 (1967)
14. Forsyth, P. J. E., C. A. Stubbington and D. Clark, *J. Inst. Metals*, **90**, 238 (1961–62)
15. McMillan, J. C. and R. M. N. Pelloux, *A.S.T.M. STP 415*, 505 (1967)
16. Richards, C. E. and T. C. Lindley, The Influence of Stress Intensity and Microstructures on Fatigue Crack Propagation in Ferritic Materials, *Engineering Fracture Mechanics*, **4**, 951–978 (1972)
17. Ritchie, R. O., The Influence of Fracture Mechanisms on Fatigue Crack Propagation, Institution of Metallurgists Spring Meeting: *The Practical Implications of Fracture Mechanisms* (1973)
18. Smallman, R. E., *Modern Physical Metallurgy*, Butterworths, London (1963)

19. Frost, N. E., K. J. Marsh and L. P. Pook, *Metal Fatigue*, Oxford Engineering Science Series (1974)
20. Sherratt, F., *Min. of Aviation*, S & T Memo, 1/66 (1966)
21. Almen, J. O. and P. H. Black, *Residual Stresses and Fatigue in Metals*, McGraw-Hill, New York (1963)
22. Love, R. J., *Symposium on Properties of Metallic Surfaces*, Institute of Metals, p. 161 (1952); *International Conference on Fatigue*, Inst. of Mech. Engrs, p. 570 (1956); Motor Industries Research Association, Re. No. 1950/9, (1950)
23. Hammond, R. A. F. and C. Williams, *Metall. Rev.*, **5**, 165 (1960)
24. Gough, J. H., *J. Inst. Metals*, **49**, 17 (1932)
25. Gould, A. J., *Int. Conf. on Fatigue*, Inst. Mech. Engrs, p. 341 (1956)
26. Evans, U. R., *Failure of Metals by Fatigue*, p. 84, Melbourne University Press (1947); *The Corrosion and Oxidation of Metals*, Edward Arnold, London (1960)
27. Gilbert, P. T., *Metall. Rev.*, **1**, 379 (1956)
28. Duquette, D. J. and H. H. Uhlig, *Trans. Am. Soc. Metals*, **62**, 839 (1968)
29. Cazaud, R., *Fatigue of Metals*, Chapman and Hall, London (1953)
30. Cazaud, R. and Pomey, G., *La Fatigue des Metaux*, Dunod, Paris (1969)
31. Waterhouse, R. B., *Fretting Corrosion*, Pergamon Press, Oxford (1972)
32. Tweedale, J. G., *Mechanical Properties of Metals*, Allen & Unwin, London (1964)
33. Creep and Fatigue in Elevated Temperature Applications, International Conference, Philadelphia (1973), Sheffield (1974)
34. ASME 1331 Code for Pressure Vessels (1974)
35. Peterson, R. E., Interpretation of Service Fractures, *Handbook of Experimental Stress Analysis* (ed. M. Hetenyi), Wiley, New York (1950)
36. Lipson, C. and Juvinall, R. C., *Handbook of Stress and Strength*, The MacMillan Co., New York (1963)
37. Dolan, T. J., Preclude Failures: A Philosophy for Materials Selection and Simulated Service Testing, *Experimental Mechanics*, **10** (1), 1 (1970)
38. Colangelo, V. J. and Heiser, F. A., *Analysis of Metallurgical Failures*, Wiley, New York (1974)
39. Failure Analysis and Prevention, *Metals Handbook*, 8th edition, Vol. 10, ASM (1975)

5

Assessment of Crack Formation Life

5.1 Introduction

In assessing the fatigue integrity of a component, it is necessary to estimate the number of cycles required to produce an engineering crack (section 3.5), and the subsequent number of cycles required to propagate the crack to a limiting condition. For the present, attention will be confined to the assessment of crack formation life, and the application of fracture mechanics to assess fatigue crack propagation will be discussed at a later stage.

5.2 Determination of Stress and Strain at a Concentration

Excluding finite element methods, two possible avenues are currently available for evaluating the strains at a concentration feature, namely the Neuber[1] and the Hardrath–Ohman (or modifed Stowell)[2] methods.

    If we consider a simulated component, for example, a tensile member containing a concentration feature such as a circular hole, the associated stress and strain distributions will depend upon the magnitude of the load, the theoretical stress concentration factor, and the material stress–strain relationship. Assuming a non-strain hardening material having perfectly elastic–perfectly plastic material characteristics, as indicated in figure 5.1, the stress distributions which would be obtained for different magnitudes of load are shown in figure 5.2. In figure 5.2 (a), the load is less than that required to cause the onset of yield, i.e. $\sigma_0 \leqslant S_p/K_\sigma$ and the stresses field is completely elastic. If the externally applied load is such that $\sigma_0 > S_p/K_\sigma$, but not so great that complete plasticity occurs, the stress distribution would be as indicated in figure 5.2 (b). Finally, for the completely plastic condition, i.e. $\sigma_0 = S_p$, the stress distribution would be as indicated in figure 5.2 (c). Concentrating on the more general case shown in figure 5.2 (b), it may be observed that such a stress distribution consists of three regions, namely:

    (1) a macroscopic plastic zone;
    (2) an elastic zone with a stress gradient; and
    (3) an elastic zone with zero (or negligible) stress gradient.

    Under completely elastic conditions (figure 5.2 (a)) the concentration of stress and strain is quite localised and will decrease rapidly at small distances away from

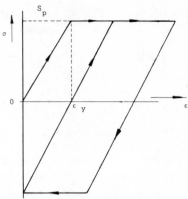

Figure 5.1 *Stress—strain relationship for an assumed material having perfectly-elastic perfectly-plastic characteristics*

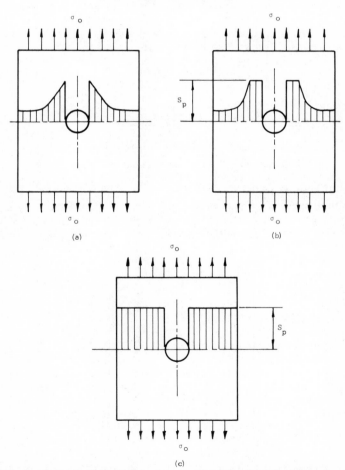

Figure 5.2 *Stress distribution obtained for a perfectly-elastic perfectly-plastic material for different magnitudes of load (a) $\sigma_0 \leqslant S_p/K_\sigma$; (b) $S_p > \sigma_0 > S_p/K_\sigma$; (c) $\sigma_0 = S_p$*

the concentration feature. The ratio of the maximum localised elastic stress or strain to the nominal or average stress or strain, based on the minimum cross-sectional area, is identical to the theoretical or geometric stress concentration factor, K_t, i.e.

$$K_t = \frac{\sigma_{max}}{\sigma_0} = \frac{\epsilon_{max}}{\epsilon_0} \tag{5.1}$$

and in this instance $K_t = K_\sigma = K_\epsilon$. The actual stress and strain distributions under elastic conditions are dependent upon the geometry and type of load, and are related to each other by the elastic modulus for the material. For the simplest case of a circular hole in a thin plate of assumed infinite width (plane stress condition), under the action of a unidirectional tensile load (figure 5.3), the stress field equations may be obtained using the equilibrium and compatibility equations in conjunction with an appropriate choice of Airy stress function. Thus, it may be shown[3,4] that at any point (r, θ) from the centre of the hole, the stresses are given by

$$\sigma_r = \frac{\sigma_0}{2}\left[\left(1 - \frac{a^2}{r^2}\right) + \left(1 - \frac{4a^2}{r^2} + \frac{3a^4}{r^4}\right)\cos 2\theta\right] \tag{5.2}$$

$$\sigma_\theta = \frac{\sigma_0}{2}\left[\left(1 + \frac{a^2}{r^2}\right) - \left(1 + \frac{3a^4}{r^4}\right)\cos 2\theta\right] \tag{5.3}$$

$$\tau_{r\theta} = \frac{-\sigma_0}{2}\left(1 + \frac{2a^2}{r^2} - \frac{3a^4}{r^4}\right)\sin 2\theta \tag{5.4}$$

these stress field equations were first derived by Kirsch[5] and are consequently known as Kirsch's solution. The maximum stress occurs at the edge of the hole, i.e. where $r = a$ and $\theta = \pi/2$, and for this condition $\sigma_r = \tau_{r\theta} = 0$, and $\sigma_{max} = 3\sigma_0$, indicating that $K_\sigma(= K_t) = 3$.

For a plate of assumed infinite width but having a finite thickness t, the results of Sternberg and Sadowsky[6] indicate that under a unidirectional axial load the maximum stress is still the tangential stress σ_θ at the edge of the hole, but that this

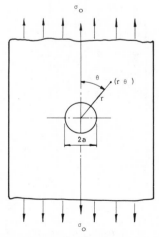

Figure 5.3 *Thin plate with a central hole under the action of a uni-directional tensile load*

value is not constant through the plate thickness. In fact, at the two outer surfaces of the plate the value for K_σ was found to be slightly less than 3, whilst at the interior of the plate, K_σ had a value slightly greater than 3. However, irrespective of the thickness to hole diameter ratio $(t/2a)$ the value for K_σ at the centre of the plate is never greater than 3.09, and at the other surface never less than 2.7, indicating that for most engineering applications the value obtained from Kirsch's solution may be reasonably assumed valid for plates of any thickness[3].

For ductile materials subjected to loads which cause the localised stresses in the vicinity of the concentration feature to exceed the yield strength, S_p, plastic flow limits the value of the stress concentration factor K_σ, although the same cannot be said for the strain concentration factor K_ϵ. When plasticity occurs, even though K_σ decreases, the plastic flow will increase the maximum localised strain with a consequential increase in K_ϵ. This increase in strain concentration may also mean that the strain distribution diminishes more slowly with distance from the concentration feature. As a consequence, the important conclusion may be reached that for the conditions involving macroscopic plasticity, the strain concentration factor may be a more important quantity than the stress concentration factor.

An iterative procedure for estimating the stress and strain distribution for the case of a circular hole in a flat plate, loaded by a unidimensional tension load which may cause plastic strain, was proposed by Stowell[7] in 1950. By assuming a set of stresses at any point (r, θ) to be represented by a modification of Kirsch's equations, Stowell derived the following equations:

$$\sigma_r = \frac{\sigma_0}{2}\left[\left(1 - \frac{a^2}{r^2}\right) + \frac{E_s}{E}\left(1 - \frac{4a^2}{r^2} + \frac{3a^4}{r^4}\right)\cos 2\theta\right] \tag{5.5}$$

$$\sigma_\theta = \frac{\sigma_0}{2}\left[\left(1 + \frac{a^2}{r^2}\right) - \frac{E_s}{E}\left(1 + \frac{3a^4}{r^4}\right)\cos 2\theta\right] \tag{5.6}$$

$$\tau_{r\theta} = -\frac{\sigma_0}{2}\frac{E_s}{E}\left(1 + \frac{2a^2}{r^2} - \frac{3a^4}{r^4}\right)\sin 2\theta \tag{5.7}$$

where E_s is the secant modulus corresponding to the actual stress–strain behaviour of the material at the point where the stresses are being described, and E is the usual elastic modulus, assumed at infinite distance from the concentration feature. The stress concentration factor, K_σ can be obtained by considering the stress at the edge of the hole, i.e. where $r = a$ and σ_θ has its maximum value when $\theta = \pi/2$. Hence,

$$K_\sigma = \frac{\sigma_{max}}{\sigma_0} = 1 + 2\left(\frac{E_s}{E}\right) \tag{5.8}$$

The corresponding equations for strain are derived[7] as

$$\epsilon_r = \frac{\sigma_0}{4E_s}\left[\left(1 - \frac{3a^2}{r^2}\right) + \frac{E_s}{E}\left(3 - \frac{8a^2}{r^2} + \frac{9a^4}{r^4}\right)\cos 2\theta\right] \tag{5.9}$$

$$\epsilon_\theta = \frac{\sigma_0}{4E_s}\left(1 + \frac{3a^2}{r^2}\right) - \frac{E_s}{E}\left(3 - \frac{4a^2}{r^2} + \frac{9a^4}{r^4}\right)\cos 2\theta \tag{5.10}$$

$$\gamma_{r\theta} = \frac{3\sigma_0}{2E}\left(1 - \frac{a^2}{r^2}\right)\left(1 + \frac{3a^2}{r^2}\right)\sin 2\theta \tag{5.11}$$

The strain concentration factor K_ϵ can be obtained in a similar manner to that used in obtaining the stress concentration factor, from which it may be shown that

$$K_\epsilon = \frac{\sigma_{\max}}{\sigma_0/E} = K_\sigma \left(\frac{E}{E_s}\right) \tag{5.12}$$

The theoretical stress and strain distributions and the stress and strain concentration factors, calculated as indicated above, have been compared by Stowell with some experimental data obtained by Griffith[8] for a wide sheet of aluminium alloy, with a central hole loaded in tension. It is concluded that the calculated stress concentration factor corresponds sufficiently well with the experimentally determined value, but calculations for the strain concentration factor are somewhat lower than the experimentally determined values. It is suggested that this apparent discrepancy is probably due to the peculiarities of the stress—strain curve which permits a slight error in stress to be enormously magnified in strain, together with the assumption that Poisson's ratio is equal to 0.5 in the plastic region.

The limitations associated with Stowell's iterative procedure are that the method can only be applied to the case of a circular hole in an assumed infinitely wide plate, and it is only applicable under conditions of static loading. A modification of Stowell's relationship for stress concentration, as represented by equation (5.8) was proposed by Hardrath and Ohman[2] in 1953, and has the more general form

$$K_\sigma = 1 + (K_t - 1)\frac{E_s}{E} \tag{5.13}$$

enabling its application to other geometries than the circular hole in an assumed infinitely wide plate; the strain concentration factor may then be estimated using equation (5.12) but in this form, the method is still only applicable to static loading conditions. More recently, experimental investigations by Javornicky[9] using photoplasticity, provide some limited experimental support for the modified Stowell relationship represented by equation (5.13).

Another method which has been suggested for determining the strain concentration factor when small plastic strains are involved is due to Neuber[1]. This relationship, originally derived for sharp notches in shear, essentially states that the theoretical elastic stress concentration factor K_t, is equal to the geometric mean of the actual stress and strain concentration factors, i.e.

$$K_t^2 = K_\sigma K_\epsilon \tag{5.14}$$

A comparison of the Neuber approximation and the modified Stowell or Hardrath—Ohman relationship has been studied by Snow[10]. It is shown that according to the Neuber approximation, the ratio of the local stress concentration factor under plastic conditions, K_σ, to the elastic stress concentration factor, K_t, is given by

$$\frac{K_\sigma}{K_t} = \left(\frac{E_s}{E}\right)^{1/2} \tag{5.15}$$

and that by the Hardrath–Ohman relationship

$$\frac{K_\sigma}{K_t} = \frac{E_s}{E} + \frac{1}{K_t} \left(1 - \frac{E_s}{E} \right) \qquad (5.16)$$

Thus, the Neuber relationship indicates that (K_σ/K_t) is dependent only on the shape of the stress–strain curve, whilst the Hardrath–Ohman equation depends both on the shape of the stress–strain curve and the geometry. For the limiting condition where (E_s/E) is equal to unity, i.e. the material remains completely elastic, both equations (5.15) and (5.16) yield the result that $K_\sigma = K_t$. When only small amounts of plasticity are involved, it is expected that there will be only small differences between equations (5.15) and (5.16), but as the amount of plastic penetration increases, the difference between the modified Stowell and the Neuber methods might be expected to increase.

Under dynamic loading conditions, the stress and strain distributions will continue to change with cycles, dependent upon the cyclic strain hardening or strain softening characteristics of the material. Whether, in fact, a crack will propagate during the period preceding a stable condition will depend upon the magnitude of the applied external load, but the important point to note at this stage is that the monotonic and cycle strain distributions will, in general be different.

Figure 5.4 illustrates diagrammatically the stress and strain distributions obtained under constant load cycling between zero and maximum, for a cyclic strain hardening material. From this illustration it may be observed that in order to obtain an accumulation of cyclic macroplasticity in the immediate vicinity of the

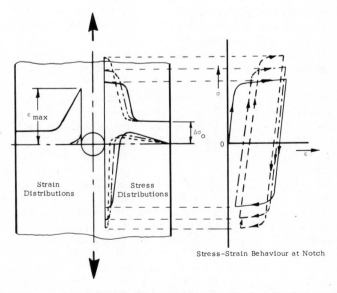

Figure 5.4 *Stress and strain distributions under constant load cycling between zero and maximum for a cyclic strain hardening material*

notch or concentration feature, it is necessary for the yield point in compression to be exceeded during the compressive part (i.e. the unloading) of the cycle. Limited experimental data provided by Crews and Hardrath[11] suggests that this does not happen except for very high nominal stresses ($\sigma_0 > 0.8 S_p$). The stress and strain redistribution during cyclic loading has also been studied by Blatherwick and Olson[12], and under constant load amplitude it was observed that both stress and strain redistributions occurred. Further, in the immediate vicinity of the concentration feature a condition approaching the constant strain hypothesis was not obtained.

A theoretical investigation has been made by McKenzie and Green[13] to assess the maximum residual stresses and strains likely to be obtained at a concentration feature typical of an aero-engine component, and 'to show to what degree zero-to-maximum strain cycling on plain specimens is likely to represent events in a notched specimen under cyclic loading'. It is concluded that cyclic strain tests on plain specimens should include tests to find the effect of mean strains of the order of 60 per cent of the alternating strain range, and that for these conditions the Neuber approximation overestimates the peak stress and strain by about 25 per cent. It is further suggested that the Neuber approximation is therefore a poor guide to the degree of plastic deformation which occurs.

In most practical designs the local plastic strains will usually be sufficiently contained to limit the plastic zone to only a small region, i.e. the ratio of plastically to elastically strained material will usually be quite small. Under these circumstances, even though the component as a whole may be subjected to constant load cycling (due for example, to start-ups and shut-downs, which may or may not involve temperature fluctuations), the material in the vicinity of a concentration feature will have a cyclic stress—strain response quite different from that of the bulk material. This local behaviour will be dependent upon a number of factors, such as the relative magnitude of plastically to elastically strained material, the strain distribution, the materials cyclic strain hardening or strain softening characteristics, and the effect of environment, more particular that of temperature. Further, even though the component may be subjected to constant amplitude cyclic loading, the material in the vicinity of the concentration feature and which is locally plastic, will experience a variation in strain range with cycles[14]. It is the behaviour of this local material, more particularly the way in which plastic strain or plastic strain energy accumulates with cycles, which governs the mechanism responsible for low cycle failure.

In order to obtain true LCF failure, it is necessary for yielding to occur on both the first and second half of each and every cycle, this phenomenon being very much dependent upon the elastic restoring forces for the bulk material acting on the local plastic region. If this condition is not obtained, and failure still occurs in the low cycle region, then according to the definition of low cycle failure which has been adopted here, failure must occur by some form of deformation[15]. One other case which requires special mention is the condition where local yielding may occur on the application of the first half cycle, but subsequent cycling reduces to elastic shakedown. In some instances of this kind, it is possible for failure to occur in a relatively few number of cycles, i.e. less than about 50 000c, but the type of failure is that associated with high cycle fatigue (HCF). Fatigue integrity for this condition can be assessed using a high cycle fatigue analysis (see Chapters 1 and 2) but the effect of a mean stress caused by the initial yielding on the first half cycle must be included in the analysis.

It has been proposed by Dawson[16] that the Hardrath–Ohman extension of Stowell's equations may be used to determine the cyclic stress and strain concentration factors, provided the cyclic stress–strain curves are used in place of the monotonic stress–strain curves. Thus, this local strain may be determined by multiplying the nominal strain by the strain concentration factor K_ϵ, and if the constant strain hypothesis is accepted an estimate of the cyclic life to crack formation can be obtained in conjunction with a constant strain fatigue curve.

The modified Stowell equation represented by equation (5.13) may be re-written as

$$\Delta\sigma = \frac{\Delta\sigma_0 \Delta\epsilon_T}{\Delta\epsilon_T - (K_t - 1)(\Delta\sigma_0/E)} \tag{5.17}$$

and the Neuber method, expressed by equation (5.15), may also be re-written in the form

$$\Delta\sigma = K_t^2 \Delta\sigma_0^2 / E\Delta\epsilon_T \tag{5.18}$$

where $\Delta\sigma$ and $\Delta\epsilon_T$ represent the maximum stress and strain ranges respectively; $\Delta\sigma_0$ is the nominal stress range; K_t is the geometric stress concentration factor; and E is the elastic modulus. Both the above methods are quicker, simpler and cheaper to use than finite element methods, but might be less accurate.

Equations (5.17) and (5.18) provide an iterative procedure for the determination of the maximum localised stresses and strains, under uniaxial stress conditions when used in conjunction with the appropriate stress–strain curve. Thus, for a known value of K_t and a particular nominal stress range $\Delta\sigma_0$, a range of values for $\Delta\epsilon_T$ may be assumed and the corresponding values for $\Delta\sigma$ calculated. Where the curve for $\Delta\epsilon_T$ against $\Delta\sigma$ intercepts the stress–strain curve for the material, this defines the maximum stresses and strains at the concentration feature. This is demonstrated in figure 5.5. Determining the cyclic strain necessitates estimating the residual strain, ϵ_{min}, after unloading, and this can be calculated by application of the Neuber or modified Stowell rule to the unloading half cycle. In this instance, of course, the material stress–strain response for the unloading half cycle must be either known or estimated. The origin for the unloading half cycle is taken as the furthest point reached on the stress–strain curve during the first half cycle of loading, as indicated in figure 5.5.

For cyclic loading under conditions of plane stress, Zwicky[17] has derived a relationship for the total strain range at a concentration feature based upon the assumption of a stable hysteresis loop being obtained. This relationship may be written as

$$\Delta\epsilon_T = \frac{2\epsilon_y'(K_t - 1)(\Delta\sigma_0/2S_p')}{1 - (\Delta\sigma_0/2S_p')} \tag{5.19}$$

where ϵ_y' and S_p' represent the cyclic yield strain and cyclic yield strength (obtained from a cyclic stress–strain curve) respectively.

Zwicky refers to an unpublished report by Sarney, in which the modified Stowell method has been extended to plane strain conditions for monotonic

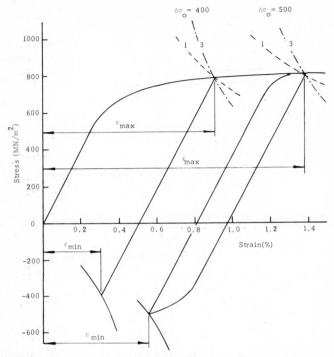

Figure 5.5 *Application of Neuber rule for plane stress for the determination of maximum and minimum local stress—strain conditions with $K_t = 3$. Nominal stresses as indicated*

loading, the secant modulus then being defined in terms of equivalent stress and strain; these latter quantities are obtained using the von Mises—Hencky criterion. Combining the cyclic loading extension of the basic modified Stowell method, equation (5.19), with the plane strain analysis of Sarney it is suggested[14] that the total strain range at a concentration feature can be estimated from the expression

$$\Delta\epsilon_T = \frac{(2\epsilon_y'/g)(K_t - 1)(\Delta\sigma_0/2S_p')}{(1/m) - (\Delta\sigma_0/2S_p')} \tag{5.20}$$

where m and g are material constants dependent upon the degree of plasticity, and are defined as:

$$m = \sqrt{(1 - \nu' + \nu'^2)} \tag{5.21}$$

$$g = m/(1 - \nu'^2) \tag{5.22}$$

$$\nu' = \tfrac{1}{2} - (\tfrac{1}{2} - \nu)\frac{E_s}{E} \tag{5.23}$$

Equation (5.20) may be written in the alternative form

$$\Delta\epsilon_T = \frac{m\Delta\sigma_0(K_t - 1)\Delta\sigma}{E(\Delta\sigma - m\Delta\sigma_0)} \tag{5.24}$$

The cyclic strain range may also be estimated in a similar manner using the Neuber rule. Rigg[18] has suggested that the assumption of elastic–perfect plastic conditions to represent the stress–strain curve may involve significant error, particularly at high strain values, and he defined the stress–strain curve by an equation of the form

$$\Delta\sigma = F\Delta\epsilon_T^b \tag{5.25}$$

Thus, for plane stress conditions Neuber's rule may be written (cf. equation (5.18))

$$\Delta\epsilon_T = (K_t^2 \Delta\sigma_0/FE)^{1/(1+b)} \tag{5.26}$$

and for plane strain conditions

$$\Delta\epsilon_T = [mg\, K_t^2(1 - \nu^2)\Delta\sigma_0^2/FE]^{1/(1+b)} \tag{5.27}$$

Accepting the assumptions inherent in the derivation of the equations presented, either the modified Stowell method or the Neuber rule may be used to estimate the strain range at critical sections in a component. However, more recent work by Duggan[19] suggests that equation (5.25) is not a suitable representation of the material stress–strain relationship.

5.3 Predicting Crack Formation Life

Since the major objective of determining the strain at a concentration feature is that of assessing fatigue integrity, it is now appropriate to develop a method which enables the number of cycles required to form an engineering crack to be estimated.

If the assumption is made that the behaviour of material at a critical region in a component can be simulated by smooth specimen behaviour, then we have a ready method for assessing crack formation. This consists essentially of using the known material cyclic behaviour, as expressed mathematically by the modified Manson–Coffin relationship, equation (3.12), with the total strain range value calculated using one of the methods of the previous section. Thus, if $\Delta\epsilon_T$ is estimated using the Neuber or modified Stowell method, the cycles to crack formation can be determined from equation (3.13), i.e.

$$N_f = \left(\frac{\Delta\epsilon_T - \Delta\epsilon_a}{\epsilon_f' - \epsilon_m}\right)^{1/\alpha_1} \tag{3.13}$$

The endurance strain range corresponding to any particular combination of mean and alternating components is estimated using a strain equivalent of the Soderberg relationship. Thus,

$$\Delta\epsilon_a = \frac{2\epsilon_{end}\epsilon_{yp}(1 - r_\sigma)}{\epsilon_{yp}(1 - r_\sigma) + \epsilon_{end}(1 + r_\sigma)} \tag{5.28}$$

where ϵ_{end} = endurance strain corresponding to zero mean strain ($= S_e/E$); ϵ_{yp} = yield strain; r_σ = stress ratio (based on local stress values).

This approach is only applicable if the fatigue curve can be accurately represented by equation (3.13). An alternative approach is to use the approximation of a linear elastic and plastic strain range against cycles relationship when plotted logarithmically, as expressed by equation (3.3), i.e.

$$\Delta \epsilon_T = \Delta \epsilon_p + \Delta \epsilon_e = C_p N_f^{\alpha 1} + C_e N_f^{\alpha 2} \tag{3.3}$$

This equation is based upon completely reversed strain cycling with zero mean stress. As already discussed (section 3.5), to allow for the effect of mean strain the Sach's modification is introduced into the plastic strain component, i.e.

$$\Delta \epsilon_p = (\epsilon_f' - \epsilon_m) N_f^{\alpha 1} \tag{5.29}$$

and in equation (3.13) the plastic strain range has been replaced by the total strain range, $\Delta \epsilon_T$. If the elastic strain range corresponding to zero mean stress is $\Delta \epsilon_{e0}$, then

$$\Delta \epsilon_{e0} = C_e N_f^{\alpha 2} \tag{5.30}$$

To allow for the effect of a mean stress the Goodman relationship may be used, i.e. in terms of strain range

$$\Delta \epsilon_e = \Delta \epsilon_{e0} \left(1 - \frac{\sigma_m}{S_u} \right) \tag{5.31}$$

Combining equations (5.30) and (5.31) and substituting into equation (3.3)

$$\Delta \epsilon_T = (\epsilon_f' - \epsilon_m) N_f^{\alpha 1} + C_e \left(1 - \frac{\sigma_m}{S_u} \right) N_f^{\alpha 2} \tag{5.32}$$

If an endurance limit exists for the material, say $\Delta \epsilon_{L0}$ at N_e cycles, then

$$C_e = \Delta \epsilon_{L0} N_e^{-\alpha 2} \tag{5.33}$$

where $\Delta \epsilon_{L0}$ is the strain range equivalent of the endurance limit for zero mean stress, i.e.

$$\Delta \epsilon_{L0} = \frac{2 S_e}{E} \tag{5.34}$$

Now it will be observed that, using an iterative procedure, equation (5.32) enables an estimate of crack formation life in a component to be predicted, if the conditions at the critical region can be determined.

The solution for strain range $\Delta \epsilon_T$, using either the Neuber or modified Stowell relationship requires an iterative procedure, applied first to the loading half cycle, and then to the unloading half cycle. It would seem reasonable to assume, for the first half cycle, that the material locally will follow the monotonic stress–strain curve. On unloading, various possibilities are likely, depending upon the magnitude of strain attained on the loading half cycle, and the overall stress ratio. Thus, the local material behaviour may be such that either:

(1) yielding in compression occurs;
(2) no yielding in compression but residual compressive stress is obtained; or
(3) residual tensile stress is achieved due to high nominal stress ratio.

These conditions are illustrated in figure 5.6. In any real component the strain distribution will be such that strain gradients exist and the volume of plastically strained material and the strain exponent may be significantly different from that of the plain test piece where the strain gradient will usually be essentially zero. A recent study by Leis *et al.*[20] suggests that the theoretical stress concentration factor, K_t, should be replaced by an experimentally determined value for K_f (termed the 'fatigue notch factor'). Investigations conducted by the authors suggest that if K_t is replaced by K_f in the relationship for estimating strain range, then

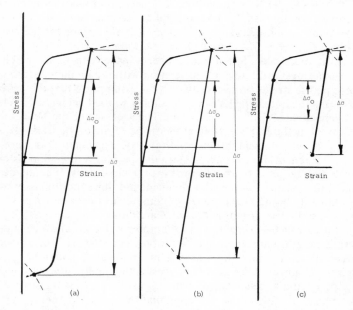

Figure 5.6 *Local material behaviour: (a) yielding in compression; (b) no yielding in compression but residual compressive stress; (c) residual tensile stress*

closer predictions to crack formation lives are obtained. However, the values for K_f were obtained using the usual definition involving notch sensitivity index (section 1.7). Once this local material behaviour has been established, the mean strain may be determined from

$$\epsilon_m = \frac{\epsilon_{max}}{2}(1 + r_\epsilon) \tag{5.35}$$

where r_ϵ is the local strain ratio. To extend the crack formation model to include the effect of bulk stress ratio, R, the nominal stress range in the calculations of total strain range is modified by including the stress ratio for the unloading half cycle. Thus, if σ_0 is the nominal stress amplitude attained on the loading half cycle, then for the unloading half cycle

$$\Delta\sigma_0 = \sigma_0(1 - R) \tag{5.36}$$

This allowance for bulk stress ratio requires further verification, and in any case it would not be expected to be applicable if time dependent influences such as creep or stress relaxation are involved.

To investigate the validity of the proposed model, experimental data for fatigue crack formation has been obtained for a variety of materials, geometries, bulk stress ratios and elevated temperatures. K_t varied from less than 2 up to about 14. Predictions for crack formation life have been made using the model represented by equation (5.32)[29]. The total strain range $\Delta \epsilon_T$ has been calculated using both the modified Stowell and the Neuber methods. For each situation it is necessary to know the service conditions (K_t and $\Delta \sigma_0$) and the appropriate material properties obtained from monotonic tests and fully reversed strain cycling fatigue tests on plain specimens (S_u, E, ν, ϵ_f', α_1 and α_2). The iterative procedure required makes a computer solution desirable. Strain measurements made using miniature electrical resistance strain gauges indicate good correlation with the predicted strains, except that for large plastic strains (in excess of 1%), the assumption that the material behaviour during the unloading half cycle is similar to the loading half cycle may lead to an overestimate of the residual strain. The accuracy of the Neuber and modified Stowell methods depends upon the shape of the cyclic stress—strain curve and the degree of plastic strain. Strain conditions are dependent upon geometry and specimen or component dimensions, and strictly speaking a three-dimensional finite element analysis is required to establish the precise conditions. If complete restraint exists at the notch, plane strain will be obtained, and this condition is approached if the ratio of notch radius (ρ) to specimen or component thickness (B) is small. As (ρ/B) increases, so the conditions approached are those of plane stress. Figure 5.7 indicates a restraint factor (F), allowing for notch configuration, to be applied to the total strain range calculated on the basis of a plane stress analysis. Figure 5.8 shows the correlation between experimental and predicated results for crack formation lives obtained on the above basis, using the modified Stowell method of calculating local material behaviour. For the majority of results the scatter is within the tolerance band of ±25%, thus suggesting that the predictive method presented has a most acceptable accuracy. Both Neuber and the modified Stowell methods overestimate actual strain range, but the modified Stowell method is generally more accurate.

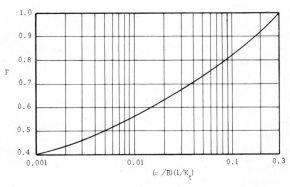

Figure 5.7 *Variation of material and restraint factor (F) with geometric parameter ($\rho/B)(1/K_t$)*
(Duggan)

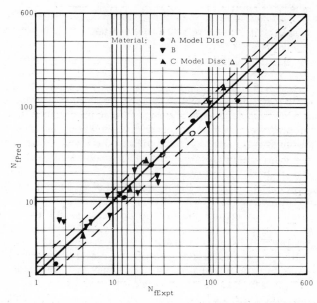

Figure 5.8 *Correlation between predicted and experimental crack formation lives (Duggan)*

5.4 Factors Influencing High Strain Fatigue

The basic Manson–Coffin relationship assumes that the mean strain which may exist has negligible effect on crack formation life. This assumption is valid[21-25], provided that the magnitude of the mean strain does not exceed the cyclic strain range, and is small compared to the fatigue ductility coefficient, i.e.

$$\frac{\epsilon_m}{\Delta \epsilon_T} \ll 1$$

$$\epsilon_m \ll \epsilon_f 1 \tag{5.37}$$

It has been shown[22,25] that a mean strain can be considered as a pre-strain by which the metal is first strain hardened and reduced in ductility; subsequent cycling is then performed on this strain hardened material at zero mean strain. The effect of this reduced ductility has been incorporated into the Manson–Coffin equation by subtracting the mean strain from the fatigue ductility coefficient (cf. equation (3.13)).

For a stable hysteresis loop, the mean strain, ϵ_m, is obtained from

$$\epsilon_m = \tfrac{1}{2}(\epsilon_{max} + \epsilon_{min})$$

$$= \frac{\epsilon_{max}}{2}(1 + R) \tag{5.38}$$

where R = strain ratio = $\dfrac{\epsilon_{min}}{\epsilon_{max}}$.

Under conditions corresponding to shakedown to elastic conditions after yielding on the first half cycle, the material is able to support a mean stress due to the initial yielding. For such instances a conventional HCF analysis may be performed; as discussed in chapters 1 and 2 the allowance for mean stress due to initial yielding must be taken into account.

A most important factor influencing high strain fatigue behaviour is that of environment, more particularly that of elevated temperature. The effect of elevated temperature and some simple aspects of creep—fatigue interactions have already been discussed in previous sections but this discussion was in the main concerned with low strain or high cycle fatigue. In the present discussion some of the effects of elevated temperature on the response of materials subjected to continuous cycling in the high strain fatigue regime will be considered.

The principal effect of increasing the temperature of an isothermal fatigue test is to decrease the endurance under strain controlled or load controlled conditions. If no significant metallurgical changes occur then the decrease in life with increasing temperature will usually be reasonably uniform. However, there may be metallurgical changes in the structure of the metal, a process known as aging. Alloys developed specially for creep resistance normally include elements which will cause precipitation during service, and such precipitation may markedly alter the properties of the material. A second metallurgical feature which may be observed is a change in the nature of the fracture, from trans- to intergranular, with increasing temperature and increasing time to failure.

Since we are mainly concerned here with that of predicting cyclic life, it is important to recognise the significance of time dependent influences in high strain fatigue at elevated temperature. Of particular significance is that of creep or stress relaxation, the effect of which is to redistribute the stress—strain profiles at notches or stress concentration features.

At elevated temperature, the strain rate or frequency effect is a most important parameter influencing life, and in order to make allowances for this and other time-dependent influences, various empirical relationships have been developed[26]. One such relationship relates stress range, plastic strain range and frequency and may be written

$$\Delta\sigma = A\Delta\epsilon_p^{n'} f^{-k_1} \tag{5.39}$$

where $\Delta\sigma$ = stress range; $\Delta\epsilon_p$ = plastic strain range; f = frequency; A, n', k_1 = constants determined from test data.

Further characterisation of high temperature fatigue behaviour is accomplished with the aid of a high temperature modification of the low temperature Manson—Coffin equation[26], i.e.

$$\Delta\epsilon_p = C_2 (N_f f^{k-1})^{\alpha} \tag{5.40}$$

from which a cyclic life equation may be obtained

$$N_f = \left(\frac{\Delta\epsilon_p}{C_2}\right)^{1/\alpha} \left(\frac{1}{f^{k-1}}\right) \tag{5.41}$$

The material constants in equations (5.39) and (5.40) can be determined from relatively short time tests and equation (5.41) can then be used to predict life when the overall time to failure is too great for normal testing to be applicable.

The subject of fatigue at elevated temperature is too complex a subject to be dealt with here in more detail, and the interested reader is referred to the proceedings of recent symposia[26,27].

5.5 Concluding Remarks

The justification for the use of any model in design is that of acceptable accuracy. In the fatigue situation the inherent scatter will always ensure some error, but to a limited extent this can be accounted for by determining reserve factors[28]. Even then, some potential unreliability will remain, since it is not possible to extrapolate to extreme values on duty and capability distribution curves with one hundred per cent confidence.

The application of the models for crack formation life predictions has been undertaken for various geometries, types of loading, materials and environment. These include sharp vee-notches in bending, both at room temperature and elevated temperature, with and without nominal stress ratios; U-notches in bending at room temperature and zero nominal stress ratio; simulated components in the form of tension plates with central holes, at room temperature; and model discs in two different materials at both room temperature and elevated temperature. Additional investigations are necessary to extend the model for use where creep—fatigue interactions are important. Once crack formation life has been assessed, the rate of fatigue crack growth may be estimated using fracture mechanics concepts, as discussed in chapter 6.

References

1. Neuber, H., Theory of Stress Concentration for Shear Strained Prismatical Bodies with Arbitrary Non-Linear Stress-Strain Law, *J. App. Mech.*, **28**, 544 (1961)
2. Hardrath, H. F. and Ohman, L., A Study of Elastic and Plastic Stress Concentration Factors due to Notches and Fillets in Flat Plates, *NACA Tech. Note 117*, Washington (1953)
3. Durelli, A. J., Phillips, E. A. and Tsao, C. H., *Introduction to the Theoretical and Experimental Analysis of Stress and Strain*, McGraw-Hill, New York (1958)
4. Sines, G., *Elasticity and Strength*, Allyn and Bacon, Boston (1969)
5. Kirsch, G., Die Theori d. Elatizatat u.d. Bedurinissed, Festigkeitslehre, *Z. Ver. deut. Ing.*, **42** (29), 799 (1898)
6. Sternberg, E. and Sadowsky, M., Three-dimensional Solution for the Stress Concentration Around a Circular Hole in a Plate of Arbitrary Thickness, *J. Appl. Mechanics* (Trans. Am. Soc. Mech. Engrs.), **16**, 27 (1949)
7. Stowell, E. Z., Stress and Strain Concentration at a Circular Hole in an Infinite Plate, *NACA Tech. Note 2073*, Langley Aeronautical Laboratory, Washington (1950)
8. Griffith, G. E., Experimental Investigation of the Effects of Plastic Flow in a Tension Panel with a Circular Hole, *NACA Tech. Note 1705*, Washington (1948)

9. Javornicky, J., Plastic Stress and Strain Concentration Factors in a Strip with a Hole or Notches Under Tension, *J. Strain Analysis*, **3** (1), 39 (1968)

10. Snow, E. W., Designing for Low Cycle Fatigue: Comparison of Neuber and Modified Stowell Rules, *Aero Stress Memorandum Internal Report ASM 2172*, Rolls-Royce, Derby (1970)

11. Crews, H. J. and Hardrath, H. F., A Study of Cyclic Plastic Stresses at a Notch Root, *Expt. Mechanics*, **6** (6), 313 (1966)

12. Blatherwick, A. A. and Olson, B. K., Stress Redistribution in Notched Specimens During Fatigue Cycling, *Expt. Mechanics*, **8** (8), 356 (1968)

13. McKenzie, R. D. and Green, J., Estimation of Stresses and Strains at a Plastic Notch, *Aero Stress Memorandum Internal Report ASM 2460*, Rolls Royce (March, 1970)

14. Duggan, T. V., Application of Fatigue Data to Design — Crack Propagation in a Simulated Component Under Cyclic Loading Conditions, *Ph.D. Thesis*, Portsmouth (1973)

15. Duggan, T. V., Low Cycle Failure Mechanisms, *Tech. Report No. F.306*, Portsmouth Polytechnic, Portsmouth (1971)

16. Dawson, R. A. T., Factors to be Considered in the Design and Operation of Turbines to Prevent Failure by Thermal Fatigue, *Proc. Conf. on Thermal and High Strain Fatigue*, Metals and Metall. Trust, London (1967)

17. Zwicky, E. E., Cyclic Strain Concentration Factors with Local Plastic Flow, *67-WA-PVP6*, Am. Soc. for Testing and Mater. (1967)

18. Rigg, G. J., Investigation into Strain Controlled Fatigue Behaviour in the Range 100—50,000 Cycles at Ambient and Elevated Temperatures, *M.Phil Thesis*, Portsmouth (1975)

19. Duggan, T. V. Unpublished Work

20. Leis, B. N., Gowda, C. V. B. and Topper, T. H., Cyclic Inelastic Deformation and the Fatigue Notch Factor, *Cyclic Stress—Strain Behaviour — Analysis, Experimentation and Failure Prediction*, ASTM STP 519, Am. Soc. for Testing and Mater. (1973)

21. Pian, T. H. H. and D'Amota, R., Low Cycle Fatigue of Notched and Un-notched Specimens of 2024 Aluminium Alloy under Axial Loading, *Wright Air Dev. WADC Tech. Note 58—27* (1958)

22. Sessler, J. G. and Weiss, V., Low Cycle Fatigue Damage in Pressure Vessel Materials, *Trans. ASME (J. Basic Engng)*, **85**, 539—546 (1953)

23. Dubuc, J., Vanasse, J. R., Biron, A. and Bazergui, A., Evaluation of Pressure Vessel Design Criteria for Effect of Mean Stress in Low Cycle Fatigue, *Proc. 1st Int. Conf. on Pressure Vessel Technology*, Delft, Netherlands, Paper 11—98 (1969)

24. Dubuc, J., Vanasse, J. R., Biron, A. and Bazergui, A., Effect of Mean Stress and Mean Strain in Low Cycle Fatigue of A517 and A201 Steels, *Trans. ASME (B)* **92** (1), 35—52 (1970)

25. Sachs, G., Gerberich, W. W., Weiss, V. and Lattorre, J. V., Low Cycle Fatigue of Pressure Vessel Materials, *Proc. Am. Soc. for Testing and Mater.*, **60**, 512—529 (1960)

26. Coffin, L. F., Fatigue at High Temperature, *Fatigue at Elevated Temperatures, ASTM STP 520*, Am. Soc. for Testing and Mater. (1973)

27. International Conference on Creep and Fatigue in Elevated Temperature Applications, Sheffield UK 1974, Inst. Mech. Engrs (1975)
28. Duggan, T. V., Quality and Reliability in Design, *The Engineering Designer*, 2 (4), 15–19 (1976)
29. Duggan, T. V. and Sabin, P., Effect of Geometry on Crack Formation, *4th Int. Conf. on Fracture*, University of Waterloo, Ontario, Canada (1977)

6

Fracture Mechanics and Fatigue Crack Propagation

6.1 Introduction

The methods of analysis proposed for assessing the integrity of components in the HCF region (discussed in Chapters 2 and 3) are concerned with total life considerations where the cyclic life spent in fatigue crack propagation may be considered small. This supposes that the life to crack formation and that to fracture are more or less coincident. In Chapter 5, the cyclic life involved in crack formation in the intermediate and LCF regions, where macroplasticity may accumulate cyclically, was discussed in some detail, but life predictions were limited to that of producing an engineering crack. Pre-existing flaws or defects have not so far been specifically considered, although the statistical nature of fatigue is, to some extent, due to inhomogeneities and structural and microstructural defects of one kind or another. The scatter in fatigue life is more prevalent in the HCF region where nucleation and crack formation takes longer, and since nucleation commences at local weaknesses or defects, this observation is not surprising.

Since all engineering materials contain flaws, and since crack formation in a component may not necessarily correspond to failure, it is obviously desirable to study the behaviour of a component or material with a pre-existing flaw or crack. Thus in general, the total life of a component will consist of:

(1) cycles to crack formation; and
(2) fatigue crack propagation life.

Many designers consider cycles to crack formation as total life, and any subsequent life spent in fatigue crack propagation as a bonus. There are many situations, however, where such an approach is unacceptable on economic grounds or other considerations, and it is therefore necessary to study the behaviour of cracks. Failure may not necessarily correspond to fracture, but to when a crack reaches a size such that the component either becomes unsafe or ceases to perform its required function. This chapter will be concerned with assessing those situations where cracks exist in components, and it is necessary to determine the significance of such cracks with regard to performance, both in terms of safety and reliability as well as satisfying function requirements. For this purpose, the fracture mechanics approach provides a very powerful tool.

Fracture mechanics is concerned exclusively with those situations involving cracks or defects, with a view to assessing whether or not such cracks will propagate under static or dynamic loading. Since propagation of a crack is controlled by the local behaviour at the crack tip, this requires a consideration of local stress, component shape and size, crack geometry and microstructure.

The impetus for the development of modern fracture mechanics stemmed from the problem of brittle fracture in high strength materials. The question then immediately springs to mind, how relevant is the mechanics of brittle fracture to fatigue failure, the basic mechanism of which has been shown to be due to cyclic plastic strain? In the case of crack propagation of a 'brittle' striation mechanism (section 4.3) then the relevance is fairly evident; however, in the case of 'ductile' striation crack propagation the question is more difficult to answer and in fact requires some understanding of the principles of fracture mechanics.

The design philosophy involved in applying fracture mechanics to fatigue problems entails first the acceptance that components can rarely be considered as flawless and secondly the establishment of whether defects will propagate and if so how rapidly and for how long?

6.2 Griffith Theory

Fracture mechanics has its origins in the work of Griffith[1] who proposed the principle of energy balance between the strain energy lost in propagating a crack and the surface energy of newly created fracture surfaces.

From early experiments it was established that the brittle fracture of solids required a critical stress normal to the cleavage plane. However, the value of these stresses was found to be of the order of a thousand times less than the theoretical strength indicated by the interatomic bonding strength, and the existence of flaws was postulated to account for this discrepancy. It was proposed by Griffith that the brittle fracture of glass occurred by the propagation of small surface flaws, and for this to occur the total energy of the system must decrease, i.e. the increase in surface energy of the two new fracture surfaces must be less than the decrease in stored elastic strain energy caused by the growth of the crack.

Griffith considered the growth of an elliptical through crack in a tensile loaded plate of an elastic material (glass) of unit thickness and infinite width (fig. 6.1). It can be shown[2] that the release of strain energy (U) is given by:

$$U = \frac{\pi \sigma^2 a^2}{E} \quad \text{(plane stress)} \tag{6.1}$$

$$U = \frac{\pi \sigma^2 a^2}{E}(1 - \nu^2) \quad \text{(plane strain)} \tag{6.2}$$

where σ = nominal applied stress; a = half crack length; E = Young's modulus; ν = Poisson's ratio.

The surface energy associated with a central through crack of length $2a$ and unit width is

$$S = 4a\gamma_s$$

where γ_s = surface energy/unit area.

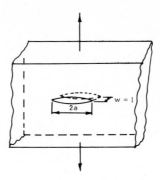

Figure 6.1 *The Griffith Model*

The variations with increasing crack length of these energies, together with the resultant energy of the system (R) are given in figure 6.2. This latter curve shows that work must be done on the plate in order to increase the crack size up to a critical value (a_{cr}), after which the crack will propagate spontaneously since its growth gives an even greater reduction in the total energy of the system.

$$R = S - U = 4a\gamma_s - \pi\sigma^2 a^2 / E \quad \text{(plane stress)} \tag{6.3}$$

$$R = S - U = 4a\gamma_s - \frac{\pi\sigma^2 a^2}{E}(1 - \nu^2) \quad \text{(plane strain)} \tag{6.4}$$

Fracture occurs when $\dfrac{dR}{da} = 0$, giving

$$\sigma_f = \left[\frac{2E\gamma_s}{\pi a}\right]^{1/2} \text{(plane stress)} \tag{6.5}$$

$$\sigma_f = \left[\frac{2E\gamma_s}{\pi(1 - \nu^2)a}\right]^{1/2} \text{(plane strain)} \tag{6.6}$$

$$\cong \left[\frac{E\gamma_s}{a}\right]^{1/2} \tag{6.7}$$

or $$\sigma_f a^{1/2} \cong (E\gamma_s)^{1/2} \tag{6.8}$$

The latter expression being a material constant.

    In practice engineering materials, even many nominally brittle ones, deform plastically to some extent and the Griffith criterion must be modified to allow for local plastic deformation at the crack tip, since this plastic deformation imparts some toughness because of the work carried out. Thus, it was proposed[3,4] that a modified equation might apply

$$\sigma_f = \left[\frac{2E(\gamma_s + \gamma_p)}{\pi a}\right]^{1/2} \text{(plane stress)} \tag{6.9}$$

where γ_p = a plastic work factor (shown experimentally to be

$$10^4 - 10^6 \times \gamma_s).$$

However, this equation was found not to be very useful since γ_p is not a material constant, but varies with crack length, strain rate and specimen geometry.

The essence of the Griffith theory, i.e. that crack propagation is an energy conversion process, has been extended by Irwin[5,6] in an attempt to find a reliable design criterion for predicting the stress at which a crack will propagate to fracture.

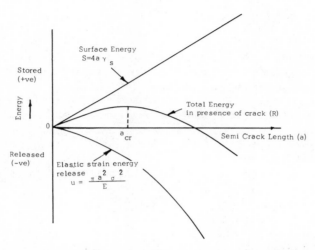

Figure 6.2 *Energy changes accompanying the growth of a Griffith crack*

This approach involves a 'crack extension force' or 'strain energy release rate', $\mathcal{G}$, which is the stored elastic energy released as the result of a crack advancing by unit area,

where

$$\mathcal{G} = \frac{dU}{d(2at)} \tag{6.10a}$$

$$= \frac{d}{d(2at)} \frac{\sigma^2 \pi a^2 t}{E} \quad \text{(plane stress)} \tag{6.10b}$$

$$= \frac{\sigma^2 \pi a}{E} \quad \text{(for unit thickness, } t) \tag{6.10c}$$

At a critical value of $\mathcal{G}$, known as the fracture toughness, the crack will propagate rapidly. For an elastic crack in an assumed infinitely wide plate of an ideal brittle material it can be shown[7] that

$$\mathcal{G}_c = 2\gamma_s$$

indicating that the Griffith and Irwin approaches lead essentially to the same result.

Whilst $\mathscr{G}$ is fairly readily found experimentally for small specimens, it is not applicable to large components and structures to which toughness measurements will be applied. For this latter situation the stress intensity factor approach has been developed, which is related to $\mathscr{G}$, and is introduced in the next section.

6.3 Linear Elastic Fracture Mechanics

Modes of Crack Growth

The three basic modes of crack surface displacement[7] which can cause crack growth, as shown in figure 6.3 are:

(I) The opening mode; where the crack faces move directly apart.
(II) The edge sliding mode; where the crack surfaces move normal to the crack front and remain in the crack plane.
(III) The shear mode; where the crack surfaces move parallel to the crack front and remain in the crack plane.

It is conventional to add the Roman numerals, I, II and III as subscripts to stress intensity factors to indicate the mode.

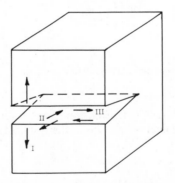

Figure 6.3 *Crack surface displacements and fracture modes (I, opening mode; II, edge sliding mode; III, shear mode)*

In fracture mechanics only the macroscopic mode of crack growth is considered, and crack growth on 45° planes (often referred to as 'shear fracture') which is a combination of Modes I and III is usually treated in calculations as if it were Mode I.

Stress Intensity Factor

Crack surfaces are stress-free boundaries adjacent to the crack tip and therefore dominate the distribution of stress in that area[7]. Remote boundaries and loading

forces affect only the intensity of the stress field at the crack tip. These fields can
be divided into three types corresponding to three basic modes of crack surface
displacement and are conveniently characterised by the stress intensity factor K
(with subscripts I, II, III denoting mode).

What then is the stress intensity factor (K)? First let us be clear what it is not. It
is not to be confused with the well known stress concentration factor (K_t), which is
the ratio of magnified to nominal stress at a notch. The stress intensity factor
characterises the elastic stress components near the crack tip, and in itself has no
physical reality.

How was this factor arrived at? Irwin[4] examined the equations of Sneddon[8] for
the elastic stress distribution around an elliptical crack in a plate, and observed that
the stress near the flaw was proportional to $\sigma\sqrt{a}$ (cf. Griffith model). Irwin then
modified the elastic stress equations to incorporate the factor $K = \sigma\sqrt{\pi a}$, which he
called the Stress Intensity Factor. This results in the following equations for the
elastic stress components near the tip of a crack (Mode I opening) in a Griffith
model situation, according to the notation shown in figure 6.4:

$$\sigma_x = \frac{K}{(2\pi r)^{1/2}} \cos\frac{\theta}{2}\left(1 - \sin\frac{\theta}{2}\sin\frac{3\theta}{2}\right) \tag{6.11a}$$

$$\sigma_y = \frac{K}{(2\pi r)^{1/2}} \cos\frac{\theta}{2}\left(1 + \sin\frac{\theta}{2}\sin\frac{3\theta}{2}\right) \tag{6.11b}$$

$$\tau_{xy} = \frac{K}{(2\pi r)^{1/2}} \sin\frac{\theta}{2}\cos\frac{\theta}{2}\cos\frac{3\theta}{2} \tag{6.11c}$$

For validity of the above expressions $\rho \ll r \ll a$ must apply (where ρ = radius of
curvature of crack tip, and a = length of edge crack or half length of centre crack).

K has the dimensions of (stress) x (length)$^{1/2}$, and therefore units of MN m$^{-3/2}$,

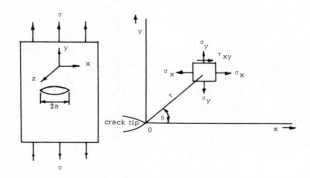

Fig. 6.4 *Elastic stress distribution in the vicinity of a Griffith crack (Mode I opening)*

and is a function of specimen dimensions, geometry and loading conditions. In general the opening mode stress intensity factor is given by:

$$K_{\mathrm{I}} = \sigma(\pi a)^{1/2} Y \tag{6.12}$$

where σ = gross tensile stress perpendicular to the crack; a = crack length (edge crack) or $\frac{1}{2}$ crack length (centre crack); Y = compliance factor.

The compliance factor, Y, is a function of geometry and loading conditions (for the Griffith model $Y = 1$), which for some simple cases may be approximated as unity, but for more complex geometries requires evaluation. Various solutions for Y have been summarised by Paris and Sih[7].

The critical value stress intensity (K_{IC}) for fast fracture (in the opening mode) is then the fracture toughness of the material, i.e. a material property, which can be found experimentally by testing specimens.

Since for the Griffith crack model:

$K = \sigma\sqrt{\pi a}$ and $\mathscr{G}$ (crack extension force) $= \dfrac{\sigma^2 \pi a}{E}$, then

$$K^2 = \sigma^2 \pi a = E\mathscr{G} \tag{6.13a}$$

i.e.
$$K^2 = E\mathscr{G} \quad \text{(in plane stress)} \tag{6.13b}$$

Practical Considerations in the Use of the Stress Intensity Factor

The above concepts of stress intensity were developed for a linear elastic solid in the form of an infinitely wide plate with simple through-crack configuration. In practice, when applying these concepts to metallic materials we need to allow for:

(1) the occurrence of plasticity;
(2) the specimen size; and
(3) other crack geometries.

(1) Crack Tip Plasticity

At the crack tip where $r = 0$, σ_y has an infinite value if equation (6.11b) is rigorously obeyed. Clearly this is physically unrealistic, since the yield stress will normally be exceeded locally prior to fracture occurring. Thus a plastically deformed zone will be formed at the crack tip, giving rise to an effective increase in crack length.

Irwin[4] has suggested that for the two-dimensional case, the plastic zone may be represented schematically by a circular boundary of radius r_p, as shown in figure 6.5, and has estimated the distance r_p based on this model using the equations for the elastic stress field. Thus, where $r = r_p$, $\theta = 0$ and $\sigma_y = S_p$ (yield stress) equation (6.11b) gives:

$$r_p = \frac{1}{2\pi} \left(\frac{K}{S_p}\right)^2 \quad \text{(plane stress)} \tag{6.14a}$$

and

$$r_p = \frac{1}{5.6\pi} \left(\frac{K}{S_p}\right)^2 \quad \text{(plane strain)} \tag{6.14b}$$

Outside the plastic zone, i.e. where $r > 2r_p$, the stress distribution is approximately the same as the elastic stress distribution around an elastic crack of half length

$$a' = a + r_p \qquad (6.15)$$

For the valid use of the stress intensity factor, based on linear elastic considerations, the relative size of the plastic zone compared to the crack length and specimen dimensions must be small. Thus r_p becomes too large when the net section stress exceeds about 0.8 of the yield stress[9]. Since it is not always possible to define the net section stress unambiguously, an alternative criterion for valid

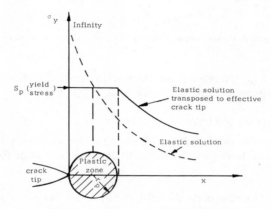

Figure 6.5 *Plastic zone at the crack tip; plane stress conditions*

plane strain stress intensity is that crack length and net section ligament shall both exceed $2.5(K_I/S_p)^2$ (Ref. 10). However, if the plastic zone size is comparable to the component dimensions, Rice[11] shows that stress intensity is no longer a valid parameter.

(2) Specimen Size

Specimen (and component) thickness is an important consideration in both determining and applying fracture toughness values. In a thick plate the yielding corresponds to plane strain conditions and plane strain plastic zone size only in the central thickness region. At the free side surfaces where σ_z falls to zero, the plastic zone size corresponds to plane stress conditions. Figure 6.6 illustrates schematically the increased size of the plastic zone towards the free surfaces resulting from the absence of restraint. Thus for thin sheet specimens the strain in the thickness direction is virtually unsuppressed and considerable plastic flow attends the cracking process. With this condition of plane stress the crack propagates through the stressed body forming an oblique shear fracture.

With increased thickness the strain in the thickness (z) direction is increasingly suppressed causing plane strain conditions (figure 6.6). Thus the magnitude of K_c, the material fracture toughness, falls rapidly with increase in specimen thickness to a

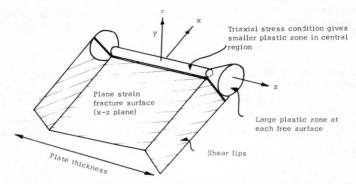

Figure 6.6 *Schematic illustration of plastic zone at a crack tip (circular zones shown are idealisations)*

minimum value, K_{IC}, the plane strain fracture toughness which is obtained for a predominantly square fracture (fig. 6.7).

(3) Variation in crack geometry

Expressions giving the value of K for a limited number of crack shapes and positions have been developed and found to have the same basic form, namely,

$$K = \sigma\sqrt{\pi a}/Q \qquad (6.16)$$

where the factor Q accounts for the effect of the crack shape and position neglecting the free surface effect discussed earlier. For a through crack in an infinite plate (Griffith model) $Q = 1$; other values for less simple crack configurations are given in the literature[7].

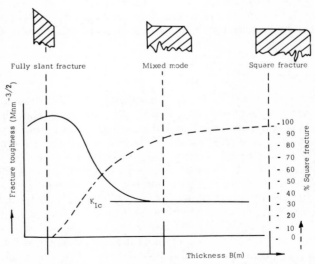

Figure 6.7 *Transition in fracture mode and fracture toughness values with increasing plate thickness. Square fracture achieved with plane strain conditions*

For the general case, then, of a flaw in a body under stress inducing mode I opening we can write

$$K_I = \sigma\sqrt{\pi(a + r_p)}\ \frac{Y}{Q} \tag{6.17}$$

where Y = compliance function; r_p = plastic zone; Q = flaw shape parameter.

Three basic types of defect are usually encountered, namely

(1) flaws which are completely submerged in the section;
(2) flaws which grow from a surface; and
(3) flaws which pass completely through the section.

Where failure takes place prior to the crack growing completely through the section the flaw, at the moment of instability, usually approximates to an ellipsoidal, semi-ellipsoidal or quadri-ellipsoidal form. Surface and embedded flaws will generally have a high degree of restraint at the crack leading edge and so plane strain conditions apply. In all cases cracks will tend to grow on a plane normal to the maximum applied (tensile) stress and so, unless some exceptional stress system exists this can be taken as the operative situation. For a submerged crack in a uniform stress field the stress intensity at any point on the boundary of an ellipsoidal crack can be calculated[12] from equation (6.17) where

$$(1/Q) = \phi^{-1}\left(\sin\beta + \frac{a^2}{b^2}\cos^2\beta\right)^{0.25} \tag{6.18}$$

where a and b are the half lengths of the minor and major axes; ϕ is the second order elliptical integral given by

$$\phi = \int_0^{\pi/2}\left[1 - \left(\frac{b^2 - a^2}{b^2}\right)\sin^2\theta\right]^{1/2}d\theta \tag{6.19}$$

and presented graphically in figure 6.8; β is the angle around the front measured

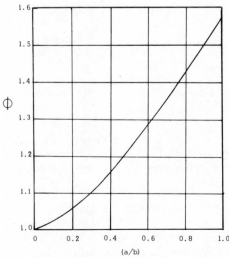

Figure 6.8 *Values for elliptic integral (equation (6.19))*

from the major axis of the ellipse. The maximum stress intensity occurs at the ends of the minor axis and the values of the modifying factor $(1/Q)$ in equation (6.18) can be presented graphically as shown in figure 6.9. Thus, for a buried crack of any aspect ratio, the value of $(1/Q)$ can be read directly from figure 6.9. Because the

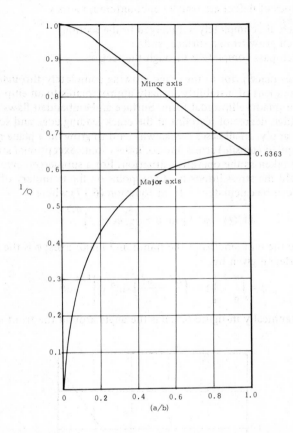

Figure 6.9 *Modification factor for elliptical crack buried in infinite body (equation (6.18))*

peak stress intensity occurs at the ends of the minor axis, a submerged defect will grow more rapidly in this direction and, unless it first becomes critical, will tend to stabilise to a circular form. Most practical cases are of this type, and equation (6.17) becomes

$$K_I = 0.636 \, \sigma(\pi a)^{1/2} Y \tag{6.20}$$

A surface crack will have a higher stress intensity factor than a corresponding buried crack, due to the fact that the restraint which controls separation of the two surfaces is reduced. This has been calculated[31] as

$$K_I = 1.1 \, \phi^{-1} \sigma(\pi a)^{1/2} Y \tag{6.21a}$$

where ϕ is given by equation (6.19) and can be read from figure 6.8. In the semi-circular form this becomes

$$K_I = 0.70 \, \sigma(\pi a)^{1/2} Y \tag{6.21b}$$

Through thickness cracks are usually only encountered in plate or sheet components and although of considerable importance for these instances, they will not be further discussed here.

From the above brief discussion, it is seen that from a knowledge of the stress field and the crack configuration, an estimate of the stress intensity factor may be made.

Fracture Toughness Testing

Fracture toughness testing is mainly limited to K_{IC} determination for the following reasons:

(1) K_{IC} is the most conservative value and is independent of specimen size;
(2) crack instability in plane stress is difficult to detect; and
(3) the large plastic deformation involved in the fracture of thin sections invalidates the analysis used for calculating K_c.

The practical determination of K_{IC} values involves the determination of the critical stress intensity factor from a knowledge of the specimen geometry, nominal stress and crack length at which fast fracture commences. In practice the onset of the latter condition may have to be approximated, i.e. the occurrence of meta-instability in the load/crack extension curve.

The specimen design and testing procedure for valid plane strain fracture toughness tests have been recommended by the British Standards Institution[14] and by Committee E24 of ASTM on Fracture Testing of Metals[15].

In brief, fracture toughness testing involves subjecting a specially designed test piece, in which a crack has been developed from a machined notch by fatigue, to either bending or tensile forces. The applied force is autographically plotted against the change in opening mode displacement measured across opposite faces of the notch during a rising force test. Crack growth is represented by an incremental increase of displacement without an increase in the applied force. The objective is to determine the force at which a given amount of crack extension takes place. This is established from the record of force-displacement in terms of a given deviation from linearity. The plane strain fracture toughness can then be calculated using the stress analysis relationship appropriate to the specific type of test piece. A diagrammatic representation of a system used for fracture toughness testing is indicated in figure 6.10.

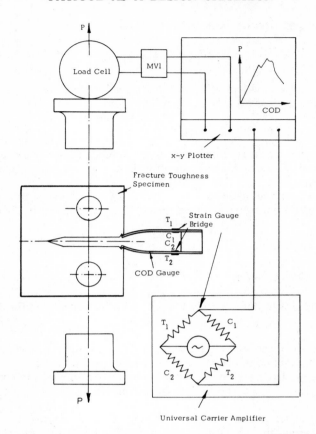

Figure 6.10 *Diagrammatic representation of system used for fracture toughness testing*

6.4 Critical Flaw Size in Fatigue

Knowledge of the plain strain fracture toughness (K_{IC}) of a material can be used to determine within reasonable accuracy, either the allowable stress intensity in the presence of a given defect or the allowable defect size in the presence of a given stress. Thus the critical crack length which when achieved during fatigue gives final fracture is predictable from:

$$a_c = \frac{1}{\pi}\left(\frac{K_{IC}Q}{\sigma Y}\right)^2 \tag{6.22}$$

Suppose we consider the case of an embedded elliptical flaw in the walls of a pressure vessel under cyclic loading. Figure 6.11 shows a graph of stress (normal to the flaw) versus flaw size (factored by Q, the flaw shape factor) for a given K_{IC} value (the fracture toughness of the material). It is evident that the pressure vessel will fail in the presence of a crack of size a_c. If the vessel is proof pressure tested to induce the stress level indicated in figure 6.11 then there can be no flaw present larger than a_1, and the minimum life of the pressure vessel is therefore determined by the crack propagation life from size a_1 to a_c.

The problem then arises of establishing the rate of crack propagation for this growth to critical size. If this is possible then it will enable an estimate to be made of the number of cycles at a given stress amplitude which will carry the crack from

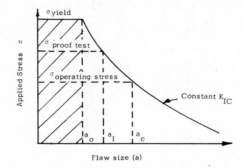

Figure 6.11 *Relationship between failure stress and flaw size*

a_1 to a_c. Furthermore, it should be possible to estimate the crack length at any given stage of fatigue and thereby gain an appreciation of the remaining life. This is a significant advance on simple S/N life testing, which only gives total failure life for a given stress amplitude.

6.5 Fatigue Crack Propagation

A fatigue crack, once formed, moves forward in a controlled manner, and the rate of crack growth depends primarily upon the amplitude of the cyclic stress or strain; the material characteristics; the geometrical configuration; and the environmental conditions. In order to guarantee that cracks which develop and grow during service do not become catastrophic, it is necessary to determine the rate of crack growth with as much reliability as possible.

During the last decade or so, considerable attention has been devoted to investigations into crack propagation relationships under cyclic loading conditions, and various laws have been proposed for the rate of fatigue crack growth with stress field, crack size and material properties. Since the elastic stress field at the tip of the crack can be fully described by the stress singularity (provided that macroscopic plasticity does not occur), it is natural that crack growth has been examined using fracture mechanics.

The rate of crack propagation is normally expressed as da/dN, where da represents the increment of crack length for an increment in number of fatigue cycles, dN. Many early attempts were made to relate da/dN to the range of applied stress, $\Delta\sigma$, and an empirical law was derived by Frost and Dugdale[16] of the form

$$\frac{\mathrm{d}a}{\mathrm{d}N} = (\Delta\sigma)^3 a/C \qquad (6.23)$$

where C is a material constant. This expression, however, makes no allowance for crack tip stress distribution, and whilst it may be applicable to simple test specimens it is not very useful for application to components.

Liu[17,18] proposed a modification to this relationship, giving the result

$$\frac{\mathrm{d}a}{\mathrm{d}N} = A(\Delta\sigma)^2 a \qquad (6.24)$$

If the loading is such that the stress varies from zero to maximum, equation (6.24) may be restated in terms of stress intensity range, i.e.

$$\frac{\mathrm{d}a}{\mathrm{d}N} = A\Delta K^2 \qquad (6.25a)$$

and since for this instance $K = K_{max}$,

$$\frac{\mathrm{d}a}{\mathrm{d}N} = AK_{max}^2 \qquad (6.25b)$$

Attempts to relate crack propagation rate to some function of stress intensity derived from the maximum stress in the cycle led to the more general result[19]

$$\frac{\mathrm{d}a}{\mathrm{d}N} = CK_{max}^n \qquad (6.26)$$

where C and n are experimentally determined constants dependent upon the material. Since equation (6.26) was originally proposed based upon investigations for a limited range of maximum stress levels and only a small variation between the minimum and maximum stress, it is hardly surprising that it lacks generality.

A model which has met with considerable success and still retains simplicity is the well known Paris relationship[20,21]

$$\frac{\mathrm{d}a}{\mathrm{d}N} = C(\Delta K_I)^n \qquad (6.27)$$

where ΔK_I is the stress intensity range, i.e.

$$\Delta K_I = K_{Imax} - K_{Imin} \qquad (6.28)$$

C and n are experimentally determined constants which are dependent upon material, mean load and environmental conditions. The average value of the exponent n was found to be approximately 4 for many materials. However, subsequently many variations in the value of n have been observed from as low as 2 to as high as 10. This variation in 'n' for steels is shown in figure 6.12[22].

Clearly if the 'Paris Law' is applicable then this greatly facilitates the prediction of growth rates in a component or structure, because relatively simple laboratory

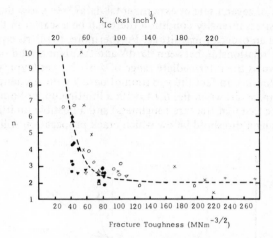

Figure 6.12 *Variation of the exponent 'n' in the Paris equation with static fracture toughness* K_{IC} *for a number of medium and high strength steels (Ritchie and Knott)*

tests can be used to obtain data which may be directly applied (via linear elastic fracture mechanics) to defects in service, provided that it is possible to analyse service stresses and stress intensity factors.

The conventional method of presenting crack propagation/stress intensity range data is in the form of a log/log plot of da/dN v. ΔK, as shown in fig. 6.13 from which the constants in the Paris equation can at first sight be readily evaluated.

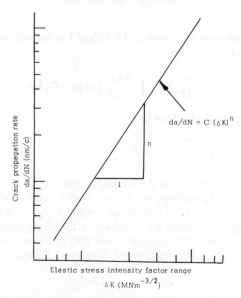

Figure 6.13 *Crack propagation plot following Paris law*

However, surveys of several sets of experimental data[21,23] show that for a given cyclic and mean stress intensity condition there can be a scatter in the da/dN v. ΔK plot of more than an order of magnitude. Furthermore, the Paris equation does not describe fully the relationship between da/dN and ΔK, being valid only for what has become known as the intermediate range of growth rates (typically 10^{-3}–10^{-1} μm/cycle). In fact the variation of da/dN with ΔK tends to be 'sigmoidal' in form, as shown in fig. 6.14, with a limiting upper bound of K_c (characterising the material fracture toughness) and a possible limiting lower bound K_{TH} (characterising a threshold below which crack propagation will not proceed).

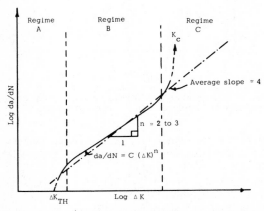

Figure 6.14 *Schematic (da/dN) versus ΔK curve for fatigue crack propagation showing the (general) sigmoidal form*

An equation developed by Broek and Schijve[24] to include the effect of mean stress can be written in the form

$$\frac{da}{dN} = C_1 \left(\frac{\Delta K_I}{1 - R} \right) \exp\left(-C_2 R \right) \tag{6.29}$$

where R is the stress ratio, defined as

$$R = \frac{\sigma_{min}}{\sigma_{max}} \tag{6.30}$$

and C_1 and C_2 are experimentally determined constants.

Each of the above relationships are subject to the criticism that they do not satisfy the limiting conditions of fast fracture when K_{max} is equal to K_{IC}, and that for non-propagating cracks when K_I is equal to a threshold level (if such a level exists). To overcome part of this criticism and also to incorporate the effect of mean stress Forman, Kearney and Engle[25] developed the equation

$$\frac{da}{dN} = \frac{C_3(\Delta K)^n}{K_c(1 - R) - \Delta K} \tag{6.31}$$

Where C_3 and n are experimentally determined constants.

Using the argument that the rate of fatigue crack propagation is probably more fundamentally related to the size of the plastic zone ahead of and in the plane of a propagating crack, Erdogan and Roberts[26,27] developed the equation

$$\frac{da}{dN} = C(K_{max})^p(\Delta K)^q \tag{6.32}$$

where C is a material constant and p and q are numerical exponents. Despite the reasonableness of the argument concerning the plastic zone size, there is little substantial evidence[28] to support the suggestion that there is a significant difference in fatigue crack propagation rates for plane stress and plane strain.

Roberts and Kibler[29] have studied the models of Forman et al. and Erdogen and Roberts, and demonstrated that for two aluminium alloys both models gave excellent results. Further, by modifying Forman's model to incorporate the ideas of Roberts and Erdogen relating to the size of the plastic zone, they have proposed a relationship which, when simplified for plane extension gives,

$$\frac{da}{dN} = \frac{CK_{max}}{K_c - K_{max}} \Delta K^2 \tag{6.33}$$

Although each of the above models are applicable for the particular set of conditions for which they were derived, they each lack generality and not one of them incorporates the condition for a non-propagating crack. In any event, the constants and exponents which appear need to be obtained for particular materials by actual crack propagation testing.

More recently investigations by Duggan[28,30] have developed a mathematical model based upon the damage which accumulates at the crack tip as a result of the irreversible process associated with plasticity. Further, the analysis recognises the resistance to fatigue crack propagation due to prior cyclic loading and the possible existence of basic threshold levels. In its original form the model may be written as

$$\frac{da}{dN} = \left(\frac{\pi}{32}\right)^{1/2\alpha} \frac{1}{\alpha} \left\{ \frac{2}{\epsilon_f' E(K_c - K_{max})} \left[\left(1 + \frac{\Delta K_{I0}}{K_c}\right) - \left(\frac{\Delta K_I}{K_c} + \frac{\Delta K_{I0}}{\Delta K_I}\right) \right] \right\}^{1/\alpha} \Delta K_I^{2/\alpha} \tag{6.34}$$

where α is the fatigue ductility exponent; ϵ_f' the fatigue ductility coefficient; E the elastic modulus; K_c the critical stress intensity factor (fracture toughness); and ΔK_{I0} is the basic threshold stress intensity range for the material. If $\Delta K_{I0}/\Delta K_I$ is small compared with unity and $\Delta K_I/K_c$, which it might be for practical values of ΔK_I, equation (6.34) reduces to (for zero stress ratio)

$$\frac{da}{dN} = \left(\frac{\pi}{32}\right)^{1/2\alpha} \frac{1}{\alpha} \left(\frac{2}{\epsilon_f' EK_c}\right)^{1/\alpha} \Delta K_I^{2/\alpha} \tag{6.35}$$

With the somewhat confusing and even contradictory array of crack growth relationships, it is important to be able to select one to meet the needs of the fracture analyst. Despite the criticisms associated with the Paris Law, equation (6.27) still remains the strongest contender at the present time, primarily because it is the most easily manipulated, but also due to the fact that a great deal of experimental data generated fits its general form over a considerable range of practical importance. However, the general form of da/dN with ΔK_I indicates that as the maximum value of K_I approaches K_{IC} the curve steepens to become infinite as fast fracture occurs. At low values of stress intensity ranges there is also a steepening, indicating that the rate of growth decreases and may eventually diminish to a point where the crack becomes non-propagating (see figure 6.14). Figure 6.15 shows typical data for an aluminium alloy RR 58 in the fully heat treated condition[13], from which it may be observed that the Paris growth law refers to only a limited section of the curve, although this section itself may be extensive.

Microstructural and other features affect the crack growth rate, but generally this influence is small compared with their affect on K_{IC}.

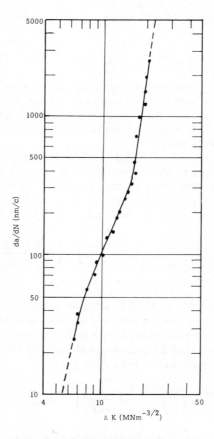

Figure 6.15 *Crack propagation data for aluminium alloy RR 58 at room temperature and R = 0.05 (McConnell[13])*

As the temperature is increased the rate at which a crack propagates is also increased, the change being mainly brought about by increased bulk strain resulting from a reduced modulus. This statement must be qualified, however, since if a material is being fatigued at temperatures towards the lower end of its ductile—brittle transition range a crack may progress in a series of unstable brittle bursts which result in apparently very high crack propagation rates. In the analysis of a quantity of unpublished data[13] it has been found that the slope n is invariably reduced and the constant C increased with increasing temperature.

At the present time there is considerable interest in the area of non-propagating cracks, since if a threshold level can be determined for a particular material and the stress distribution in a component can be accurately described, this enables the maximum permissible flaw size or defect that can be tolerated without the crack growing to be determined, provided the stress intensity factor can be estimated.

There is a considerable literature citing experimental evidence of cracks which exist under cyclic stress conditions without propagation. Frost and his co-workers[31] have determined limiting values for C_p in the equation

$$\sigma_0^3 a = C_p \tag{6.36}$$

where σ_0 is the highest pulsating stress which will not propagate a crack of size a. More recently, this work has been re-presented in terms of stress intensity[23] and it has been shown that, with some exceptions, the value of the limiting condition for non-propagating cracks is approximately related to Young's modulus. In situations where cyclic lives can be measured in giga cycles (10^9) and above, weld defects may still propagate[32] at rates which may be interpreted as zero in normal fatigue tests.

6.6 Factors Influencing Crack Propagation Rate

Whilst it is clear that the Paris law describes the fatigue crack propagation behaviour of metals in many situations, there is distinct deviation from the relationship at both low and high stress intensity ranges. Therefore in attempting to assess the effect of influencing factors on crack propagation, it simplifies matters to some extent to subdivide the da/dN versus ΔK curve into three regimes, A, B and C, as shown in figure 6.14 and consider the effects for each regime.

The parameters which may be modified in the crack propagation diagram are: the magnitude of da/dN for a specific ΔK value; the slope of the curve in the middle regime ('n' in the Paris equation (6.27)) and the ΔK values for the transitions from A to B to C regimes (i.e. departure from the Paris law). For a given material the factors which appear most likely to modify the curve for a given material are: the mean stress intensity attained, usually varied in terms of the load ratio (R) where $R = K_{min}/K_{max}$; the fracture toughness of the material (K_{IC}); the material thickness, which governs whether plane stress or plane strain conditions are dominant; the strain rate; material microstructure; temperature; and corrosion.

Richards, Lindley and Ritchie[33] have surveyed the effect of microstructure, mean (and maximum) stress intensity and specimen thickness, for a wide range of steels and some high strength aluminium alloys, particularly in the range of medium and high growth rates, and their findings are summarised in figure 6.16.

This survey shows that in the intermediate crack growth rate regime (regime B) the basic mechanism of crack growth is striation formation (see section 4.4), but with some contribution from 'static mode' crack advance giving increased growth rates and introducing 'scatter' into the da/dN versus ΔK results. Fundamentally the striation mechanism of crack growth appears to be very dependent on the reversed plastic zone size induced at the crack tip and therefore dependent on ΔK (for a given material, temperature and environment). Thus it was found that in regime B, there was little influence of microstructure, mean stress, dilute environment and thickness on crack growth rate, particularly for ferritic steels in both parent and weld metal conditions[34].

At high values of $\Delta K/K_{max}$ (resulting in a high mean value of K), departure from striation growth to include the "static mode" mechanisms was found to lead to higher growth rates (regime C). In this situation a large influence of microstructure, mean stress and thickness is evident[33]. Monotonic ('static') modes in this regime give an increase in the exponent 'n' in the Paris equation, and where bursts of brittle cracking occur this results in large accelerations in da/dN leading to much increased 'n' values.

Since in low toughness materials the crack propagation will tend to be more dominated by brittle fracture modes, then an increase in 'n' would be expected with decrease in fracture toughness (K_{IC}), which in practice is observed (figure 6.12). In fact steep slopes ('n' greater than about 3) occur almost entirely in materials of low fracture toughness (K_{IC} less than approximately 60 MN m$^{-3/2}$) (Ref. 33).

In the low growth rate regime (A), where ΔK is typically less than about 15 MN m$^{-3/2}$, as ΔK is progressively lowered the crack growth rate diminishes until a threshold value of ΔK is reached (ΔK_{TH}) below which fatigue cracks remain dormant or a very low growth rate is still sustained but which in practical engineering terms may be insignificant. There is evidence, particularly from the

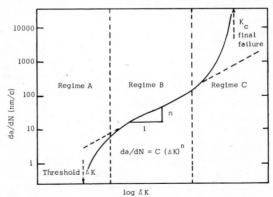

Figure 6.16 *Summary diagram showing the primary fracture mechanisms associated with the sigmoidal variation of fatigue crack propagation rate (da/dN) with range of stress intensity (ΔK) (Richards, Lindley and Ritchie) (Regime A: Non-continuum mechanisms, large influence of (1) microstructure, (2) mean stress, (3) environment; Regime B: Continuum mechanism (striation growth), little influence of (1) microstructure, (2) mean stress, (3) dilute environment, (4) thickness; Regime C: Static mode mechanisms (cleavage, intergranular and fibrous), large influence of (1) microstructure, (2) mean stress, (3) thickness; little influence of (4) environment)*

work of Beevers *et al.*[35,36] on low alloy steels and titanium alloys, in this regime that growth becomes sensitive to the maximum value of stress intensity (K_{max}); also that under these conditions the influence of the environment becomes more marked and that the microstructure influences growth where the microstructure size (e.g. grain size) is greater than the plastic zone size.

Fatigue tests on aero-engine compressor blades in RR 58 aluminium alloy have shown the *S-N* curve to be still falling at lives in excess of 10^9 cycles. It is, therefore, suggested that for design purposes, non-propagating data should be generally confined to situations where lives are not likely to be in excess of 10^7 cycles. For most engineering materials corresponding to this condition, a value of K/E for non-propagation can be taken as 4×10^{-5} m$^{1/2}$.

From the available evidence in published work there is no apparent significant effect of speed of loading on the growth rate characteristics of a material over the range 0.25–100 Hz, in the absence of a corrosive environment, although some slight increase in da/dN may occur at lower loading speeds. Frost *et al.*[37] suggest that if at low frequency the crack tip opening for a given external load is greater than at a higher frequency, then the length of fresh surface created in each cycle will be greater, giving a greater da/dN.

There is a paucity of information in the literature on the effect of temperature on crack growth rate and this is rarely expressed in terms of da/dN versus ΔK plots. Also there is a problem of separating the effects of temperature from those of environment, when tests are carried out in air. In general the effect of increased temperature in air would be expected to give an increased da/dN, but provided the temperature is still low enough for linear elastic fracture mechanics to apply the general form of the da/dN versus ΔK curve should remain unaltered, as shown in figure 6.17[38]. The actual increase in crack growth rate purely due to thermal

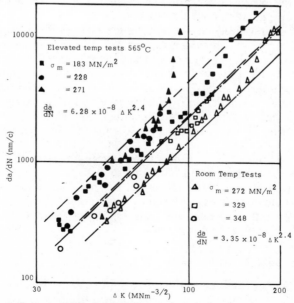

Figure 6.17 *Fatigue crack growth characteristics of 1 Cr-Mo-V low alloy steel at RT and 565°C (Ellison and Walton)*

degradation of material properties (e.g. due to increased bulk strain resulting from a reduced modulus) is probably relatively small as compared with the accelerated atmospheric effects. This importance of atmospheric corrosion is suggested by work carried out on low cycle high strain fatigue (where a large proportion of life is made up of crack propagation) in air and high vaccum[39,40].

This atmospheric effect will be both frequency and wave-form dependent (i.e. dwell-time dependent). Thus low frequency cycling would be expected to be more susceptible to atmospheric attack, and dwell at peak stress would allow both longer time for corrosion and creep deformation. The problem of fatigue–creep interaction in crack propagation is highly complex and the subject of a number of current research programmes[41].

The effect of environment on fatigue crack propagation is one of the major remaining areas requiring investigation, and some considerable activity has developed in very recent years[34–36]. Much data has been obtained for fatigue crack growth in air and for stress-corrosion crack growth in aggressive environments, using similar crack growth rate measurement techniques and in both cases applying linear elastic fracture mechanics. In the latter work attempts have usually been made to establish a threshold stress intensity (K_{ISCC}) for immunity to crack growth under constant load in a specific corrodent. Attempts have been made to correlate the interaction of fatigue and stress corrosion by treating each as an independent crack growth process. This has become known as the 'superposition model' and requires that the same stress-corrosion mechanism should operate under both static and dynamic conditions and is generally restricted to frequencies below 10 Hz[42,43]. However, predictions of growth rates obtained using this model are not considered to be sufficiently accurate compared with observed data[44]. Also fatigue growth at stress intensities below K_{ISCC} has been observed to be accelerated by the environment[45], thus no lower bound to corrosion acceleration of fatigue crack propagation can be defined.

Nicholson[44] has reviewed current work on this fatigue–stress corrosion interaction and carried out environmental testing for both conditions of crack growth. It was found that environments of sodium chloride solution, hydrogen and hydrogen sulphide can influence crack growth even when the maximum stress intensity (K_{max}) is below the K_{ISCC} value determined under static load tests. However, it was also found that in environments of hydrogen and hydrogen sulphide, high frequency fatigue cycling can effectively reduce the crack velocity to below that occurring under static load, even when the minimum stress intensity is above K_{ISCC}.

The problem of predicting the interaction between fatigue and stress corrosion in crack propagation is exacerbated further by the effects of frequency, load ratio (controlling K_{max}) and possibly temperature. This has resulted in the setting up of some elaborate (and expensive) test facilities to attempt to closely simulate the specific service conditions of some high duty applications as for example, in water cooled and sodium cooled[46] nuclear reactor plant.

6.7 Assessing Crack Propagation Life

The appreciation of component failure is frequently clouded by concepts based on experience with cylindrical test pieces subjected to a fixed mode of stressing for a measured number of cycles, or hours, before separating into two pieces to provide a point on a graph. Fatigue, however, is a continuous progression through the stages of initiation, propagation and final fracture.

A new component will generally be produced in an uncracked state, and after a finite number of cycles, a fatigue crack may form and propagate. Early fatigue damage detection has occupied much effort and has not met with universal success. Further, the prediction of fatigue crack initiation and the subsequent growth to an engineering size crack, typically up to 0.5 mm in length, is difficult to achieve with a high confidence level. The life of a crack, once formed, is more predictable, and its rate of growth and critical size are quantities which can be calculated with some degree of certainty, although influences such as geometry, material, manufacturing history, operating environment and so on often affect the behaviour and reduce the degree of certainty associated with the assessment.

The most common cause of failure in a component is probably due to an unsuspected stress level in excess of that predicted for the service condition under which the component is required to operate. A less common cause of failure is due to material properties which are below those anticipated when the design calculations were undertaken. This may be due to a basic material deficiency, to the manufacturing or processing route, or to the adverse effects of the environment.

Practical interest in fracture, potential or real, falls into two major areas, namely in the diagnosis of failures which have already occurred, and in the prognosis of situations in order that measures may be taken to avoid failure. The prediction and prevention of failure is somewhat more difficult than that of failure diagnosis because the problems involved and the measure of success are not always apparent.

The shape of a fatigue crack is primarily determined by the stress field in which it grows, but so far there is little published data available on which to base predictions. However, based on service records for rotating compressor components in which cracks had been detected, all the cracks were semi-elliptical surface flaws. Once initiated, it would appear safe to infer that when cracks are formed they retain their early shape throughout unless other factors intervene.

Subcritical crack growth under single level cyclic conditions, or where blocks of load level applications are regular, can be calculated by integrating the appropriate crack growth law which applies. Thus, based on the Paris law, equation (6.27),

$$\int_{N_1}^{N_2} dN = C^{-1} \int_{a_0}^{a_2} \left[\sigma(\pi a)^{1/2} \frac{Y}{Q} \right]^{-n} da \qquad (6.37)$$

where equation (6.17) has been used to replace ΔK_I. If the relationship between σ and crack length, a, is known, this can be incorporated into equation (6.37). For the situation where the crack is growing in an essentially constant stress field, then

$$(N_2 - N_1) = N_c = C^{-1} \left(\frac{Y}{Q} \sigma \pi^{1/2} \right)^{-n} \int_{a_0}^{a_2} (a^{-n/2}) da$$

therefore

$$N_c = \frac{C^{-1}(\sigma \pi^{1/2})^{-n}}{(1 - n/2)} \left(\frac{Y}{Q} \right)^{-n} [a_2^{1-n/2} - a_0^{1-n/2}] \qquad (6.38)$$

A suitable initial size of a_0 for the crack has to be chosen. Examination of the fracture surface may reveal some initial defect such as a slag inclusion or a quenching crack. In the absence of such a source, an arbitrary size will need to be decided. The minimum defect size detectable by current non-destructive testing techniques provides one such size.

Before fatigue integrity is proved, a number of questions need to be asked. It should, of course, be emphasised that no design is totally proof against failure, but risks can be minimised. The most important questions to be considered when assessing a design from the point of view of fatigue are:

(1) Where will the crack start?
(2) Will the crack propagate?
(3) What is the cracked life?
(4) Will the crack be catastrophic?
(5) Can the design be improved?
(6) Can the material be improved?
(7) Can manufacture be improved?

It is not the purpose of this chapter to consider these questions in detail, but in order to assess the cracked life of components in service, an attempt should be made to answer them.

6.8 Fatigue Crack Propagation Testing and Analysis

In common with all other methods of component integrity assessment, fracture mechanics necessitates basic material data. In general the data required consists of fatigue crack propagation rates for increasing stress intensity; the stress corresponding to unstable or critical conditions (i.e. the fracture toughness); and, in exceptional circumstances the conditions relating to a basic threshold level below which a crack will not propagate.

Crack propagation data under fatigue conditions requires testing standard specimens under cyclic loading and monitoring crack length with cycles. Differentiation of the crack length against cycles curves will give crack growth rate $(\mathrm{d}a/\mathrm{d}N)$ and if the equation for stress intensity with crack length is known, then the $(\mathrm{d}a/\mathrm{d}N)$ against (ΔK) curve may be produced. Specimens may be tested in tension or bending, and under conditions of displacement or load control. Testing which enables the stress intensity at the crack tip to be maintained constant is very appealing, since the resultant crack length against cycles relationship should be a straight line of slope equal to the crack growth rate.

This method of testing may be achieved either by decreasing the load with increase in crack length or by testing a specially designed specimen such that as the crack grows the stress intensity is maintained constant[47]. In any case, whatever the test procedure adopted, the crack length needs to be monitored with cycles.

A common test procedure used to obtain basic fatigue crack propagation data is that of using a single edge notched (SEN) test piece in simple bending. A typical test rig and associated equipment is shown in figure 6.18.

Essentially, the rig consists of a cantilever beam clamped securely at one end,

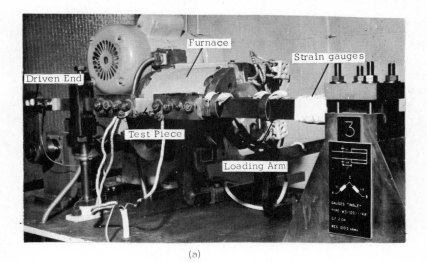

(a)

(b)

Figure 6.18 *Typical fatigue test rig for obtaining fatigue crack propagation*

whilst the free end is deflected through a fixed amplitude by an offset circular rotating cam. The test piece is clamped securely at each end to form an integral part of the cantilever, with the notch in the upper position and at the centre of the cantilever, as shown diagrammatically in figure 6.19. The height of the cantilever can be adjusted using shims at the fixed end, and the amplitude obtainable at the free end can be varied by selecting a cam position with an appropriate throw. In this way, the rig provides facilities for adjusting the initial stress and stress intensity to give appropriate values, and also permits the inclusion of different stress ratios. The drive is provided by a Carter variable speed gearbox.

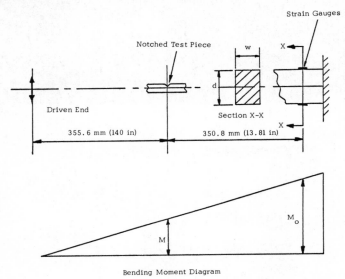

Figure 6.19 *Diagrammatic representation of bending test rig*

The fixed end of the cantilever is completely insulated in order to utilise the electric potential in measuring crack length. This method is fully discussed elsewhere[28,48], and is based upon the principle that if a crack propagates through an electric potential field, the potential measured across the crack will vary as the crack length varies.

Strain gauges positioned at the fixed end of the cantilever enable the strain to be experimentally determined at this point, and consequently the bending moment at the notch can be determined as indicated in figure 6.19.

Heating of the specimen for elevated temperature tests is achieved by a simple circular furnace arrangement incorporating cartridge heaters, and temperature is measured and controlled using thermocouples in conjunction with a temperature controller.

Before testing, each specimen is examined for surface cracks using dye penetrants. The specimens are carefully measured, the dimensions recorded, and hardness tests are made. The test facility is calibrated to ensure accurate temperature control throughout a test.

For each test the voltage across the propagating crack (related to crack length), the strain in the moment arm (related to bending moment at the crack) and the temperature of the test specimen are continuously monitored and transferred onto punched tape using the data acquisition system[48] shown diagrammatically in figure 6.20.

Although from a fundamental point of view it is desirable to test at a constant strain rate, during fatigue crack propagation the bending moment decreases with

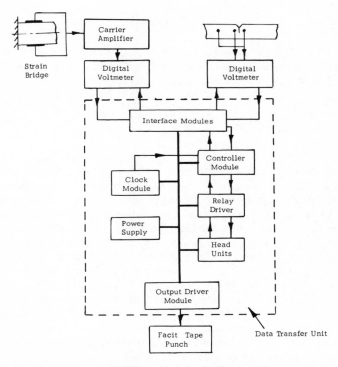

Figure 6.20 *Diagrammatic representation of data acquisition system*

cycles, and the situation is complicated. Consequently, each test is normally conducted at a constant frequency, usually 8.3 Hz (500 vib/min), thus enabling the testing to be completed as quickly as possible. Attempts are made to avoid compressive stresses at the crack tip so that the crack does not close up during a test, but even with positive nominal stress ratios crack closure may occur.

The displacement control characteristic of the rig may cause some complications at elevated temperature, due to the time-dependent parameter being significant and giving rise to some stress relaxation. This complication can be aggravated if the stress relaxation is sufficient to cause considerable crack closure due to the compressive stresses created at the crack tip.

From the records of strain in the moment arm and electric potential across the crack, the variation of bending moment and crack length with cycles can be determined using the appropriate calibrations, and these results can be used to produce fatigue crack growth rates (da/dN) as a function of stress intensity range (ΔK_I).

For the case of a single edge notched specimen loaded in pure bending, the stress intensity factor corresponding to any particular crack length may be determined from the equation[10,40,50].

$$K_I = \frac{6M}{BW^{3/2}} Y_0 \qquad (6.39)$$

where Y_0 is a modified compliance function. Using a boundary collocation procedure[50] it has been shown that the modified compliance function may be represented by the polynomial equation

$$Y_0 = 1.99 \left(\frac{a}{W}\right)^{1/2} - 2.47 \left(\frac{a}{W}\right)^{3/2} + 12.97 \left(\frac{a}{W}\right)^{5/2} - 23.17 \left(\frac{a}{W}\right)^{7/2} + 24.8 \left(\frac{a}{W}\right)^{9/2}$$

$$(6.40)$$

From the data collected by the data acquisition system and recorded onto punched tape, the crack length corresponding to any number of cycles can be determined, and the differential of this curve gives the fatigue crack propagation rate (da/dN). At corresponding points, the bending moment M and the modified compliance function Y_0 can be calculated, and hence the stress intensity can be determined using equation (6.39). In this way, the rate of fatigue crack propagation (da/dN) can be obtained as a function of stress intensity range (ΔK_I).

Obtaining the crack length and the corresponding stress intensity range with good accuracy is reasonably straightforward, but this is not necessarily true for the determination of fatigue crack growth rates. It is necessary to analyse discrete data relating the instantaneous cycle, say N_i, to the corresponding crack length, say a_i. Unfortunately, it is sometimes difficult to obtain a smooth steadily increasing curve for the da/dN graph, and different numerical treatments of the same experimental data have yielded significantly different solutions.

The commonly used method of drawing tangents to the curve of crack length against cycles may not be a particularly accurate and consistent technique for obtaining crack propagation rates, although if a Paris relationship is accepted it provides a quick and useful approximation.

A more acceptable method which has been applied successfully[28] is that of simple finite differences. This method involves taking successive data points (a_i, N_i), (a_{i+1}, N_{i+1}) and making linear interpolations to estimate the gradient at the mid-point. This technique can result in erratic gradients due to the inherent scatter in the experimental data, but with the availability of a large number of data points this error is reduced.

Recent studies into the different methods of analysing fatigue crack propagation data[48] suggest that the most accurate approach is to utilise a method suggested by Smith[51] which is based upon a set of functions proposed by McCartney and

Gale[52]. Briefly, it is assumed that the experimental data can be represented by 'm' pairs of co-ordinates, such that

$$y_i = A_1 + \frac{A_2}{x_i} + A_3 \ln x \qquad (6.41)$$

where
$$y_i = N_i - N_1 \quad \text{and} \quad x_i = \frac{a_i}{a_1}.$$

The coefficients in equation (6.41) are obtained from a least squares analysis, and thus

$$\frac{dy_i}{dx_i} = \frac{-A_2}{x_i^2} + \frac{A_3}{x_i} = \frac{A_3 x_i - A_2}{x_i^2} \qquad (6.42)$$

The inverse of this function enables the crack growth rate to be obtained, i.e.

$$\frac{dx_i}{dy_i} = \frac{x_i^2}{A_3 x_i - A_2} \qquad (6.43)$$

But
$$\frac{dx_i}{dy_i} = \frac{d(a_i/a_1)}{d(n_i - N_1)} = \frac{1}{a_1}\left(\frac{da_i}{dN_i}\right)$$

therefore
$$\frac{da_i}{dN_i} = a_1 \frac{dx_i}{dy_i} = \frac{a_1 x_i^2}{A_3 x_i - A_2} \qquad (6.44)$$

Since it is optimistic to expect equation (6.41) to fit the whole range of experimental data, several equations (typically three) are calculated with overlapping sections. To arrive at a position to commence evaluation of the numerical analysis of the data to produce crack growth rates involves considerable programming coupled with some interesting problems. The method is fully discussed elsewhere[48].

Which of the above briefly discussed methods is used to analyse the experimental data depends upon a number of factors, such as the range being covered, the time available, the accuracy required and the amount of data collected. As already observed, if considerable data is available and the range of stress intensity or crack growth rates is extensive, the so-called Smith analysis[51] provides the most acceptable method.

Figures 6.21–6.24 show the correlation between experimental data and theoretical predictions obtained using the Duggan simplified model represented by equation (6.35). In this instance the experimental data was analysed using the method of finite differences. Figure 6.25 shows some experimental fatigue crack propagation data for the heat affected zone of a welded joint in aluminium alloy LM13[55]; this experimental data was analysed using the Smith analysis already referred to.

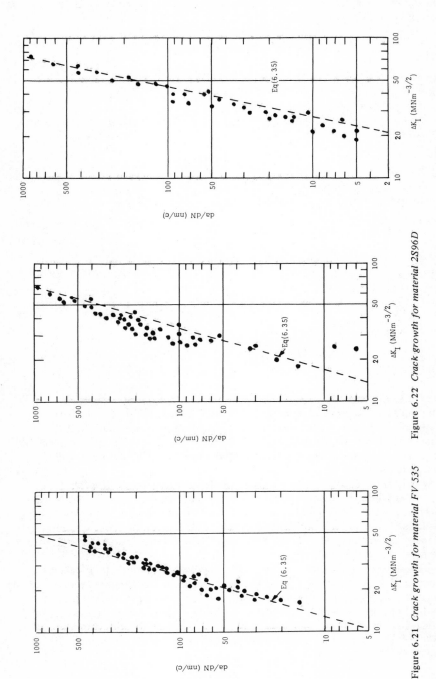

Figure 6.23 Crack growth for material Inco 901

Figure 6.22 Crack growth for material 2S96D

Figure 6.21 Crack growth for material FV 535

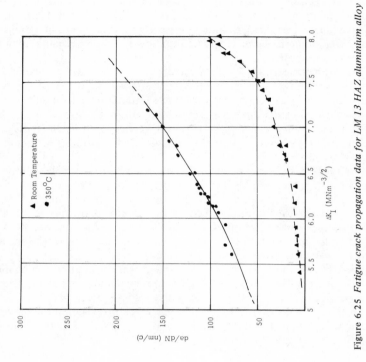

Figure 6.25 *Fatigue crack propagation data for LM 13 HAZ aluminium alloy*

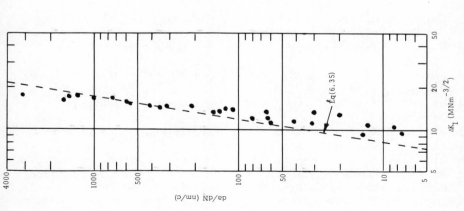

Figure 6.24 *Crack growth for material Ferrotic 'C'*

6.9 Concluding Remarks

This chapter has considered the fracture mechanics approach to fatigue. Whilst the discussions are by no means exhaustive, it is believed that sufficient has been presented to enable a basic understanding of fracture mechanics to be obtained. Further, an attempt has been made to establish how fracture mechanics may be used in a design situation.

The major problem associated with the application of fracture mechanics to design situations is that of determining the stress intensity factor. This is dependent upon load, geometry and crack length, and its determination can be quite complex. Another complication is that, in a real component, it is not always clear just how a crack will start and subsequently grow. However, service failures can be diagnosed using fracture mechanics concepts, because under failed conditions the crack geometry can usually be defined.

Stress intensity relations can be determined using a boundary collocation procedure[53] or, where a theoretical stress analysis is not possible or requires verification, using an experimental procedure suggested by Irwin and Kies[54]. However, if the stress distribution can be reasonably estimated, and the initial shape and size of the crack can be measured or approximated, making reasonable assumptions an estimate of cyclic life involved in fatigue crack propagation can be made using the methods suggested in this chapter.

An estimate of the number of cycles involved in crack formation can be made as discussed in chapter 5. If this crack formation, corresponding to the definition of an engineering crack (i.e. about 0.5 mm long and 0.15 mm deep for a surface crack), is taken as the initial defect size, then the crack propagation life may be assessed. The total fatigue life for the component will then be the sum of the crack formation life and the crack propagation life.

References

1. Griffith, A. A., Phenomenon of Rupture and Flaws in Solids, *Phil. Trans. Roy. Soc. Series A,* **221**, 163–198 (1920)
2. Knott, J. F., *Fundamentals of Fracture Mechanics*, Butterworths, London (1973)
3. Felbeck, D. J. and Orowan, E., Experiments on Brittle Fracture of Steel Plates, *Welding J. Res. Suppl.,* **34**, 5705–5755 (1955)
4. Irwin, G. R., Fracture, *Encyclopedia of Physics*, Vol. 6, p. 551, Springer, Heidelberg (1958)
5. Irwin, G. R., Onset of Fast Crack Propagation in High Strength Steels and Aluminium Alloys, *NRL Report 4763* (1956)
6. Irwin, G. R., Kies, J. A. and Smith, H. L., Fracture Strength Relative to Onset and Arrest of Crack Propagation, *Proc. Amer. Soc. Test. Mater.* **58**, 640 (1958)
7. Paris, P. C. and Sih, G. C., Stress Analysis of Cracks, *Fracture Toughness Testing, ASTM STP 381*, Am. Soc. for Testing and Mater., Philadelphia (1967)
8. Sneddon, I. N., Distribution of Stress in the Neighbourhood of a Crack in an

Elastic Solid, *Proc. Roy. Soc.*, **A187**, 229 (1946), 229–60, October 22, 1946
9. Erdogan, F., Fracture Problems in Composite Materials, *Engng Fracture Mechanics*, **4** (4), 811 (1972)
10. Brown, W. F. and Srawley, J. E., Plain Strain Testing of High Strength Metallic Materials, *ASTM STP 410*, Am. Soc. for Testing and Mater., Philadelphia (1966)
11. Rice, J. R., Mechanics of Crack Tip Deformation and Extension by Fatigue, *Fatigue Crack Propagation ASTM STP 415*, Am. Soc. for Testing and Mater. (1967)
12. Irwin, G. R., The Crack Extension Force for a Part-Through Crack in a Plate, *Trans. Am. Soc. mech. Engrs, J. app. Mech.*, **84** (4), 651–4 (1963)
13. McConnell, L. D., Cranfield Institute of Technology, Private Communication
14. B.S. 5447: 1977, Methods for Plane Strain Fracture Toughness, British Standards Institution, London (1977)
15. Brown, W. F. (Ed.), Tentative Method of Test for Plane Strain Fracture Toughness of Metallic Materials, *Review of Developments in Plane Strain Fracture Toughness Testing*, ASTM STP 463, Am. Soc. for Testing and Mater. (1970)
16. Frost, N. E. and Dugdale, D. S., The Propagation of Fatigue Cracks in Sheet Specimens, *J. Mech. Phys. of Solids*, **6**, 92–110 (1958)
17. Liu, H. W., Crack Propagation in Thin Metal Shut Under Repeated Loading, *Trans. Am. Soc. mech. Engrs*, Series D, 83 (1961), 23
18. Liu, H. W., Fatigue Crack Propagation and Applied Stress Range – An Energy Approach, *Trans. Am. Soc. mech. Engrs*, Series D, 85 (1963), 116
19. Christensen, R. H. and Harmon, M. B., Limitations of Fatigue Crack Research in the Design of Flight Vehicle Structures, *Fatigue Crack Propagation ASTM STP 415*, Am. Soc. for Testing and Mater. (1967)
20. Paris, P. C., The Fracture Mechanics Approach to Fatigue – An Interdisciplinary Approach, *Proc. 10th Sagamore Army Materials Research Conference*, Syracuse University Press, New York (1964)
21. Paris, P. C. and Erdogan, F., A Critical Analysis of Crack Propagation Laws, *J. basic Engng*, ASME Series D, 85 (1963), 528
22. Ritchie, R. O. and Knott, J. F., Mechanisms of Fatigue Crack Growth in Low Alloy Steel, *Acta Met.*, **21**, 639 (1973)
23. Frost, N. E., Pook, L. P. and Dentan, K., A Fracture Mechanics Analysis of Fatigue Crack Growth Data for Various Materials, *Engng fracture Mechanics*, **3** (2) 109 (1971)
24. Broek, D. and Schijve, J., The Influence of Mean Stress on the Propagation of Cracks in Aluminium Alloy Sheets, *NRL – Tech. Rep. No. M211* (1963)
25. Forman, R. G., Kearney, V. E. and Engle, R. M., Numerical Analysis of Crack Propagation in Cyclic Loaded Structures, *J. basic Engng*, **89**, 549 (1967)
26. Roberts, R. and Erdogan, F., The Effect of Mean Stress on Fatigue Crack Propagation in Plates Under Extension and Bending, *Trans. Am. Soc. Mech. Engrs, J. basic Engng*, **89**, 885 (1967)
27. Erdogan, F. and Roberts, R., A Comparative Study of Crack Propagation in Plates Under Extension and Bending, *Proc. Int. Conf. on Fracture*, Sendian, Japan (1965)

28. Duggan, T. V., Application of Fatigue Data to Design — Crack Propagation in a Simulated Component Under Cyclic Loading Conditions, *Ph.D. Thesis*, Portsmouth (1973)
29. Roberts, R. and Kibler, J. J., Some Aspects of Fatigue Crack Propagation, *Engng Fracture Mechanics*, **2**, 243 (1971)
30. Duggan, T. V., A Theory for Fatigue Crack Propagation, *Symposium on Mechanical Behaviour of Materials*, Kyoto, Japan (1974)
31. Frost, N. E. and Greenan, A. F., Cyclic Stress Required to Propagate Edge Cracks in Eight Materials, *J. mech. Engng Sci.*, **6** (3), 203–210 (1954)
32. Rizk, W. and Seymour, D. F., Investigations into the Failures of Gas Circulators and Circuit Components at Hinkley Point Nuclear Power Stations, *Proc. Inst. mech. Engrs*, **179** (1 No. 21), 627 (1964–5)
33. Richards, C. E., Lindley, T. C. and Ritchie, R. O., The Mechanics and Mechanisms of Fatigue Crack Growth in Metals, *Conf. on The Mechanics and Physics of Fracture*, Churchill College, Cambridge Jan. (1975)
34. Richards, C. E. and Lindley, T. C., The Influence of Stress Intensity and Microstructure on Fatigue Crack Propagation in Ferritic Materials, *Engng fracture Mechanics*, **4** (4) 1972
35. Cooke, R. J., Irving, P. E., Booth, G. S. and Beevers, C. J., The Slow Fatigue Crack Growth and Threshold Behaviour of a Medium Carbon Alloy Steel in Air and Vacuum, *Eng. Fract. Mech.*, **7** (1) (1975)
36. Irving, P. E., Robinson, J. L., Cooke, R. J. and Beevers, C. J., *Conf. on The Mechanics and Mechanisms of Fatigue Crack Growth in Metals*, Churchill College, Cambridge, January (1975)
37. Frost, N. E., Marsh, K. J., Pook, L. P., *Metal Fatigue*, Oxford Engineering Science Series, p. 264 (1974)
38. Ellison, E. G. and Walton, D., Fatigue, Creep and Cyclic Crack Propagation in a 1Cr-Mo-V Steel, *International Conf. on Creep and Fatigue in Elevated Temperature Applications*, Philadelphia, Sept. 1973 and Sheffield, April, London (1975)
39. Coffin, L. F., The Effect of High Vacuum on the Low Cycle Fatigue Law, *Metals Transactions*, **3**, 1777 (1972)
40. Coffin, L. F., The Effect of Vacuum on the High Temperature Low Cycle Fatigue Behaviour of Structural Metals, *Corrosion Fatigue; Chemistry, Mechanics and Microstructure*, NACE Houston, NACE-2, p. 590–600 (1972)
41. *International Conference on Creep and Fatigue in Elevated Temperature Applications*, Philadelphia, September, 1973 and Sheffield, April (1974)
42. Wei, R. P. and Landes, J. D., Correlation between Sustained-load and Fatigue Crack Growth in High Strength Steels, *Mater. Res. and Stds.*, **9** (7), 25–28 (1969)
43. Gerberich, W. W., Birat, J. P. and Zackay, V. F., Superposition Model for Environmentally Assisted Fatigue Crack Propagation, *International Conference on Corrosion Fatigue*, Stons, Conn., 14th June (1971)
44. Nicholson, C. E., Influence of Mean Stress and Environment on Crack Growth, British Steel Corporation, *Conf. on Mechanics and Mechanisms of Crack Growth*, Cambridge University, 4–6 April (1973)
45. Barsorn, J. M., Corrosion Fatigue Crack Propagation below K_{ISSC}, *Engng Fracture Mech.*, **3**, 15–25 (1971)

46. Weeks, R. W., Kassner, T. F. and Weins, J. J., Influence of Sodium and Radiation on the Creep and Fatigue of Fast Reactor Components, *International Conference on Creep and Fatigue In Elevated Temperature Applications*, Philadelphia, September (1973) and Sheffield, April (1974)
47. Radon, J. C. and Culver, L. E., Intergranular Stress Corrosion and Corrosion Fatigue of Aluminium Alloy RR 58 – a Fracture Mechanics Approach, *Symposium on Mechanical Behaviour of Materials*, Kyoto, Japan (1974)
48. Brooks, D. A. and Duggan, T. V., A Study of Fatigue Crack Propagation in a $2\frac{1}{2}$ per cent Nickel–Chromium–Molybdenum Direct Hardening Steel, *Tech. Report No. F.314*, Portsmouth Polytechnic (1974)
49. Walker, E. F. and May, M. J., Compliance Functions for Various Types of Test Specimen Geometry, *BISRA Report No. MG/E/307/67*, London (1967)
50. Gross, B. and Scrawley, J. E., Stress Intensity Factors for Single Edge Notch Specimens in Bending or Combined Bending and Torsion by Boundary Collocation of a Stress Function, Tech. Note D-2603, NASA (1965)
51. Smith, R. A., The Determination of Fatigue Crack Growth Rates from Experimental Data, *Int. J. Fracture*, **9**, 452–5 (1973)
52. McCartney, L. N. and Gale, B., A Generalised Theory of Fatigue Crack Propagation, *Proc. Roy. Soc.*, A, 322–41, London (1971)
53. Scrawley, J. E. and Gross, B., Stress Intensity Factors for Crackline Loaded Edge-crack Specimens, *NASA Tech. No. TD-D-3820* (1967)
54. Irwin, G. R. and Kies, J. A., Critical Energy Rate Analysis of Fracture Strength, *Welding J. Res. Suppl.*, **33**, 1935 (1954)
55. Duggan, T. V., Assessment of Fatigue Crack Propagation in a Welded Diesel Engine Piston, *Tech. Rep. No. F.317*, Portsmouth Polytechnic (1975)

Appendix

Tutorial Examples

1. Obtain a generalised (approximate) fatigue curve for steel having an ultimate tensile strength of 830 MN m^{-2}, and a yield strength of 580 MN m^{-2}. Assume a polished specimen of 8 mm diameter, reversed bending.

2. Estimate the life which might be expected from the above specimen when subjected to the following combinations of stress:

$$\sigma_m = 414 \text{ MN m}^{-2}$$
$$\sigma_{alt} = \pm 330 \text{ MN m}^{-2}$$

Compare the results obtained using

(a)	Gerber's parabola;	(b)	Goodman line;
(c)	Soderberg's method;	(d)	Heywood's method.

(Ans. (a) $N = 4.2 \times 10^5$, (b) $N = 3.6 \times 10^3$, (c) $N = 0$, (d) $N = 8.1 \times 10^3$)

3. Seven specimens of an experimental part are fatigue tested at a standard load, with failure occurring after 20 000, 25 000, 35 000, 36 000, 42 000, 55 000 and 70 000 cycles.
(a) Show graphically how these data correspond to a normal and to a log-normal distribution.
(b) Selecting the better of the above distributions, determine the sample mean and the sample standard deviation.

(Ans. (b) 4.57, 0.187)

4. The ranked fatigue data for tests on a British Standard steel, subjected to reversed bending, is indicated in the following table (table A.1).

Using a statistical analysis, determine probability fatigue curves corresponding to 10, 20, 30, 40 and 50 per cent probability of failure. Investigate the way in which the stress level influences the type of distribution, i.e. normal or log-normal.

148

Table A.1 Ranked fatigue data (life $\times 10^5$ cycles)

Principal stress (MN m^{-2})					
Specimen no.	695	618	540	436	386
1	0.0646	0.1350	0.3240	1.3602	2.396
2	0.0760	0.1487	0.3928	1.3951	3.817
3	0.0875	0.1647	0.4239	1.4458	3.979
4	0.088	0.1656	0.4258	1.4466	4.036
5	0.0887	0.1718	0.4312	1.5418	4.285
6	0.092	0.1840	0.4333	1.8785	4.679
7	0.0938	0.1876	0.4471	1.9032	5.404
8	0.116	0.2000	0.4657	1.9209	5.909
9	0.1223	0.2571	0.5751	2.0620	5.995

5. The component shown in figure A.1 is subjected to an axial load which fluctuates between P_{min} = 178 kN and P_{max} = 712 kN. If the part is to be machined from steel having an ultimate tensile strength of 1068 MN m^{-2}, estimate

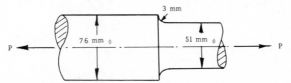

Figure A.1 *Axial member subjected to fluctuating tension*

the life which may be expected from the part using the generalised (approximate) fatigue curve. A reserve factor of 1.5 is desirable, and this should be applied to both the mean and alternating components.

6. The link shown in figure A.2 is to be made from nickel steel, A1S1 2330, water quenched and tempered at 540°C to give the following mechanical properties (based on size 12 mm diameter).

$$S_u = 930 \text{ MN m}^{-2}: S_p = 870 \text{ MN m}^{-2}; \text{ BHN} = 277$$

The load F is $\pm$25 kN and is repeated indefinitely. Determine the required diameter

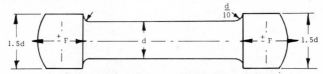

Figure A.2 *Link subjected to reversed axial loading*

'd' to give a reserve factor of 1.4 based on infinite life, for a machined surface. What would be the percentage saving in weight if the surface were mirror polished? What would be the percentage increase in weight if the surface were as forged?

7. Figure A.3 shows the cover plate design for a cylindrical pressure vessel, the internal pressure of which varies between 0 and 10 MN m^{-2}. A copper armoured asbestos gasket is employed, having a thickness of 6 mm, and a modulus of

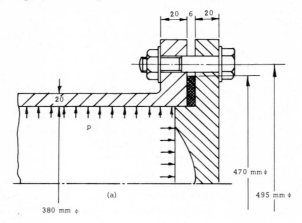

(a)

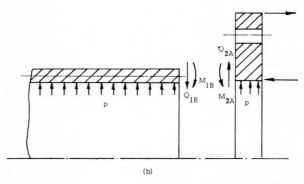

(b)

Figure A.3 *Simple cover plate design for cylindrical pressure vessel*

elasticity of 110 000 MN m^{-2}. No temperature increase is involved, so that creep relaxation of the bolts need not be considered.

From physical considerations and availability of bolts, the following bolt specification has been established.

Number of bolts	8 equally spaced
Type of thread	16 mm (M16 x 1.5)
Length of bolt	72 mm
Material properties	$S_u = 860$ MN m^{-2}
	$S_p = 690$ MN m^{-2}

A general reserve factor of 1.5 is desirable, to be applied to both the mean and alternating components of stress. Estimate the number of pressure cycles which the bolts can safely withstand.

8. A helical compression spring is to be designed for a high speed cam mechanism which will have significant higher order Fourier components in the forcing frequency. A dynamic analysis of the mechanism indicates that the spring must exert a force which varies between 110 N maximum and 54 N minimum, corresponding to an axial elongation or compression of 3.5 mm. From experience of the occurrence of fatigue failure of springs on similar mechanisms, using a statistical approach it has been established that a reserve factor of 1.3 is desirable. The feasible mean coil diameter is limited to 13 mm, due to space restrictions, and the number of active coils should not be less than 3, due to practical considerations.

On the basis of maximising the natural frequency for the lowest mode of internal longitudinal vibrations, an optimum design study has been conducted, from which the following specification has been drawn up:

Material: AS5 Music Wire; S_u = 2013 MN m^{-2}
S_{sp} = 724 MN m^{-2}
Diameter of wire: 2 mm
Mean coil diameter: 12 mm
Spring to be shot-peened (especially inner surface)
No notches or scratches permissible.

Assess the suitability of the spring from a fatigue standpoint.

9. A thin walled tube has an internal diameter of 100 mm. It is subjected to a static internal pressure to 6 MN m^{-2} and a static tensile pull of 260 kN. Determine the required wall thickness according to:

(a) Maximum principal stress; (b) maximum shear stress; (c) distortion energy or von Mises criterion.

If the material is steel having the following mechanical properties, which theory of failure is likely to be the most applicable?

$$S_u = 690 \text{ MN m}^{-2}; S_p = 518 \text{ MN m}^{-2}$$

Use a general reserve factor of 1.5.

10. Repeat example 9 for a static internal pressure of 6 MN m^{-2}, but with a fluctuating tensile pull which varies from 45 kN to 260 kN. Infinite life is required.

11. A shaft from a gear unit is to transmit a steady torque M_t of 475 N m, and at the critical point, a completely reversed bending moment of ±710 N m. As part of a feasibility study, it is necessary to obtain an estimate of the required shaft diameter. Using the equivalent damage concept, and assuming reasonable values for RF and K_t, estimate a reasonable shaft diameter to give a life in excess of 10^6 cycles. The material initially selected is En 22, oil quenched at 200°C, to give the following properties:

$$S_u = 1944 \text{ MN m}^{-2}; S_p = 1200 \text{ MN m}^{-2}$$

12. A circular shaft is subjected to a twisting moment which varies from a value of 112 N m to 224 N m and to a fluctuating bending moment that varies from a value of −112 N m to 224 N m. An axial pull varying from 1 kN to 2 kN also acts. A feasibility study has tentatively suggested the geometry at the critical section to be as indicated in figure A.4. If a life in excess of 10^5 cycles is required, what is the usage factor for the shaft? The shaft is to be machined from steel having a tensile strength of 1378 MN m^{-2}.

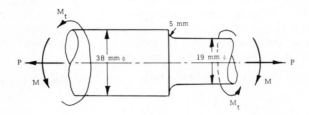

Figure A.4 *Shaft subjected to combined loading*

13. An initial analysis for a multicylinder marine diesel engine has indicated that at a critical section of crankshaft the resultant twisting moment is likely to vary from zero to 120 000 N m in phase with a variation in resultant bending moment from 10 000 N m to 50 000 N m. A diagrammatic representation of the critical section is shown in figure A.5.

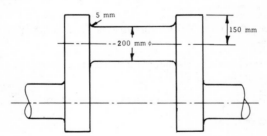

Figure A.5 *Critical section of crankshaft*

Making any necessary and reasonable assumptions, continue with the feasibility study to the stage where an appropriate material may be specified to withstand fatigue failure, based upon infinite life conditions. In order to account for the stochastic nature of fatigue and manufacturing variations, the load capability of the crankshaft is conservatively estimated to have a tolerance band of ±30%; a failure rate of zero is desirable.

14. The proposed scheme for a torsion spring is shown in figure A.6.

As the applied force *P* is completely reversed between the limits ±5 kN, the lever must rotate through an angle of about, but not exceeding ±5° from the position

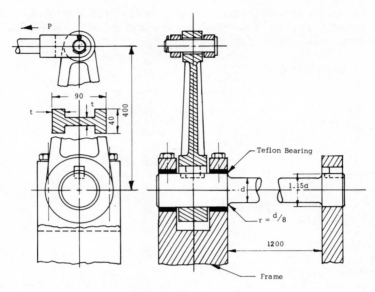

Figure A.6 *Proposed scheme for torsion spring (dimensions in mm)*

shown if it is to function correctly. Available materials for the shaft are listed in table A.2 and based on the decision to produce the lever as a casting; the alternative materials for the lever are listed in table A.3. Space restrictions are such that the overall dimensions indicated in figure A.6 must not be exceeded. Variation in peak load is not expected to be significant for this application, although due to manufacturing variations and the stochastic nature of fatigue, the load capability of the lever is conservatively estimated to have a tolerance band of ±30% and that of the shaft of ±20%.

Table A.2 Available shaft materials

Material	S_u (MN m^{-2})	S_p (MN m^{-2})	E (MN m^{-2})	G (MN m^{-2})	S_e (MN m^{-2})	Cost index per unit mass
En. 8 (R)	700	460			350	1.0
En. 17	1000	700			500	1.75
En. 22	770	500			380	1.80
En. 24(V)	1000	800	20 600	82 600	500	2.0
(W)	1080	900			520	2.2
(X)	1160	970			560	2.4
(Y)	1240	1050			600	2.5
(Z)	1544	1310			750	3.0

Making reasonable assumptions, carry out a design study in sufficient detail to specify geometry and material for both the lever and the shaft, for a design life in excess of one million cycles.

Table A.3 Available cast materials

Material	S_u (MN m^{-2})	S_e (MN m^{-2})	Cost index per unit mass
Grey Cast Irons			
BS.1452			
Grade 10	154	50.82	1.0
Grade 12	185	61.1	1.3
Grade 14	216	71.3	1.4
Grade 17	262	86.5	2.0
Malleable Cast Irons			
Whitehart			
BS.309			
w 22/4	340	136	4.0
w 24/8	370	148	5.0

15. A prototype design for the rear axle of a racing car is shown in figure A.7. As the car tranverses a circular test track the anticipated forces applied to the rear tyre are as indicated. The force F is the force supported by the outside wheel, being assumed that it corresponds to one third of the total vehicle mass. The side and

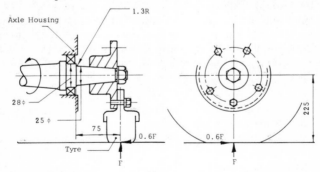

Figure A.7 *Prototype design for racing car rear axle (dimensions in mm)*

tangential forces are each assumed 0.6F, limited by an assumed coefficient of friction, based upon experimental data, of 0.85. The axle is to be manufactured from En 24(Z) nickel–chromium–molybdenum steel, having an ultimate tensile strength of 1540 MN m^{-2} and a yield strength of 1300 MN m^{-2}; all critical surfaces are to have a ground finish.

(a) Making any necessary decisions and reasonable assumptions, carry out a design study in sufficient detail to enable an estimate of the vehicle mass which would permit infinite life under continuous operation at the conditions given.

(b) What are the main limitations associated with the design study carried out in (a)?

(c) What simple design change would substantially increase the fatigue strength of the axle?

16. The inlet stub of a relief valve has the shape and dimensions indicated in figure A.8 and is to be manufactured from Monel alloy 400. The anticipated loading history is estimated to be as follows:

$$1000 \text{ cycles zero to } 15 \text{ MN m}^{-2} \text{ at } 300°\text{C}$$
$$2000 \text{ cycles zero to } 10 \text{ MN m}^{-2} \text{ at } 300°\text{C}$$
$$1500 \text{ cycles zero to } 7 \text{ MN m}^{-2} \text{ at } 300°\text{C}$$

For this combination of material and temperature, creep is not significant, and the mechanical properties of Monel alloy 400, at the temperature of interest, are as follows:

$S_u = 525 \text{ MN m}^{-2}$; $S_p = 147 \text{ MN m}^{-2}$; $E = 174\,400 \text{ MN m}^{-2}$; RA = 70%.

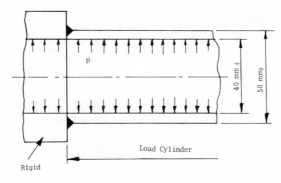

Figure A.8 *Inlet stub of relief valve*

The estimated variation capability, based upon previous design experience, is assessed as ±20% and the anticipated loading history includes an allowance for possible variations in imposed duty. Making reasonable assumptions, assess the integrity of the valve stub at the weld, on the basis of zero failure rate being a necessary objective.

17. A horizontal cylindrical vessel for a chemical processing plant (figure A.9) has an internal diameter of 1200 mm, and it is proposed to produce the vessel from Monel alloy 400, using 8 mm thick plate. The anticipated working life of the vessel is estimated as 10^5 zero to maximum pressure cycles at a constant working temperature of 300°C; for this combination of material and temperature, creep is not significant.

The vessel is to be stiffened by means of a steel reinforcing ring positioned at mid-length, and having the dimensions shown in the scrap view of figure A.9. If the

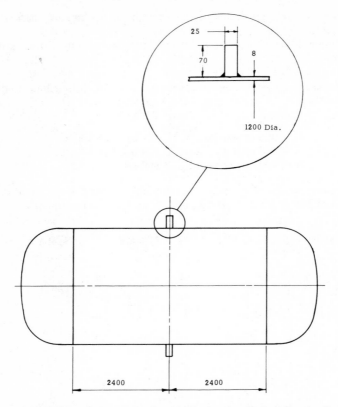

Figure A.9 *Overall dimensions and details of stiffening ring for horizontal cylindrical vessel (dimensions in mm)*

maximum pressure intensity is 1.4 MN m^{-2}, assess the fatigue integrity of the vessel at the stiffening ring welds.

The mechanical properties of Monel alloy 400, at the temperature of interest, are as follows:

S_u = 525 MN m^{-2}; S_p = 147 MN m^{-2}; E = 174 400 MN m^{-2}; RA = 70%.

18. A component in a high pressure restraining system is to be designed with a high level of confidence in integrity, but with cost as an important parameter. Based on physical limitations regarding space, and convenience of manufacture, the tentative decision has been to produce the component in the form of a thin cylinder, with a maximum permissible external diameter of 500 mm.

The system is to withstand an internal pressure intensity of 15 MN m^{-2}, and in the steady state operating condition heat transfer considerations suggest an average wall temperature of 450°C. Based on the condition of continuous operation, the creep strain rate should not exceed 0.5% in three years.

Making reasonable assumptions, carry out a design feasibility study for the pressure restraining component, with creep as the design criterion; suggest a suitable combination of wall thickness and material. Available materials, with details of creep data and relative costs are listed in table A.4. The scatter in creep data may be assumed to be within ±20%, and the variation in internal pressure is not expected to be in excess of ±10%; this latter variation may be considered insignificant as far as fatigue is concerned.

Table A.4 Available pressure vessel materials

Material	Mechanical and creep properties at 450°C stress MN m^{-2}			Cost index per unit mass
	S_p	1% creep strain		
		10 000 h	100 000 h	
B.S. 1501 : 151				
23B	114	90	60	1.0
26B	130	97	66	1.1
Colmo 900				
Grade 29	174	193	148	1.2
Grade 31	186	193	148	1.3
Colmo 950	180	278	196	1.5
Colmo 1000	220	310	191	1.75

19. A single edge-notched specimen, having the geometry shown in figure A.10, is subjected to essentially zero to maximum cyclic loading in bending. The maximum bending moment is 200 N m and the actual nominal stress ratio $R = 0.06$. The specimen thickness in 6 mm.

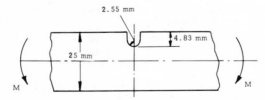

Figure A.10 *Single edged notched specimen*

Fatigue testing of the specimen at room temperature indicates a crack formation life of 97 kcycles. The material is 2S96D, having the following mechanical properties at 20°C.

$$S_u = 928 \text{ MN m}^{-2}; S_p = 768 \text{ MN m}^{-2}, \text{ i.e. } 0.1\% \text{ Proof}; E = 208\,200 \text{ MN m}^{-2};$$
$$\text{RA} = 67.5\%; S_e = 464 \text{ MN m}^{-2}.$$

Construct a low cycle fatigue curve for the material, and use this in conjunction with a modified Stowell approach, allowing for the effect of local mean stress, to estimate by prediction the crack formation life. The stress–strain curve for the material is shown in figure A.11.

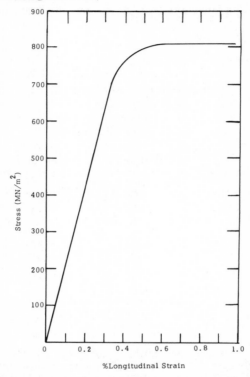

%Longitudinal Strain

Figure A.11 *Stress–strain curve for 2S96-D at room temperature*

20. In a particular component, inspection techniques have indicated a surface defect. As an approximation, this defect is semi-elliptical in shape with an aspect ratio (i.e. the ratio of half minor to half major axis) of 0.481, the major axis being in the direction of crack growth. A finite element analysis of the stress field indicates a stress distribution in the direction of crack growth of the form

$$\sigma_{max} = 9.5 - 0.423\, a\, (MN\ m^{-2}); \ \sigma_{min} = 0$$

where a is the depth from the outer surface measured in mm. Crack propagation data obtained from experimental studies suggests a crack growth rate equation of the form

$$\frac{da}{dN} = 3.425 \times 10^{-10} \Delta K^{3.13}\ m/\text{cycle}$$

Determine whether or not the crack will grow, and if it will produce a graph of crack length with cycles.

Author Index

Subject Index